T0229490

Machine Learning for Knowledge Discovery with R

Machine Learning for Knowledge Discovery with R

Methodologies for Modeling, Inference, and Prediction

Kao-Tai Tsai

Frontier Informatics Services
Bristol Myers Squibb
Adjunct Professor, Jiann-Ping Hsu College of
Public Health, Georgia Southern University

CRC Press
Taylor & Francis Group
Boca Raton London New York

CRC Press is an imprint of the
Taylor & Francis Group, an **informa** business
A CHAPMAN & HALL BOOK

First edition published 2022

by CRC Press
6000 Broken Sound Parkway NW, Suite 300, Boca Raton, FL 33487-2742

and by CRC Press
2 Park Square, Milton Park, Abingdon, Oxon, OX14 4RN

© 2022 Kao-Tai Tsai

CRC Press is an imprint of Taylor & Francis Group, LLC

The right of Kao-Tai Tsai to be identified as author of this work has been asserted by him in accordance with sections 77 and 78 of the Copyright, Designs and Patents Act 1988.

ISBN: 978-1-032-06536-6 (hbk)
ISBN: 978-1-032-07159-6 (pbk)
ISBN: 978-1-003-20568-5 (ebk)

Typeset in CMR10
by KnowledgeWorks Global Ltd.

DOI: 10.1201/9781003205685

Dedicated this book to:

My Parents

My Family

My Teachers

and My Brothers

Contents

Preface

Data analysis has a long history. Actually, the Kingdom of Sumer (Babylonia c. 4500-1900 BC) is a famous example of where initial records of written data analysis have been found. Through human history, data analysis had become the foundation for scientific discoveries, advances of medicine, and government policies.

Due to the advances of sciences, business, and computing technologies, etc., data of medium to huge sizes have become ubiquitous. In addition, with the emphasis of "everything evidence-based," how to extract valid, useful, and actionable information is more critical than ever and heavily depends on well-done data analysis. In a broader scope of data analysis, analyst needs to consider the relevance of population the data was from and whether it fits the objectives of the analysis (ref. Mallows). In the ICSA presentation (2010), I also emphasized the importance of experimental design, rigorous data collection plan to ensure the data quality, the plan to implement the analytical findings, and potential revision of the original questions. This cycle can often repeat itself a few times.

Data analysis should not be just a pure mechanical exercise following mathematical models or canned programs, instead, data analysis very often involves tedious and careful effort to examine the nature of data, including its distribution, homogeneity, correlation, outliers, and other complex internal structure, etc. before any effort of estimation or modeling, especially in large data sets (ref. Chapter 1).

Analysis of large real-world data sets is often very complex and requires a team with multi-discipline. Therefore, data analysts must look to a heavy emphasis on judgment about the subject-matter experience of the field the data being considered, and the broad experience of the analytical techniques have worked out in a variety fields of application and its properties (ref. Tukey). In addition, rather than spend great effort to pursue the "optimal/best" solutions, the time can be better used in real research (ref. Kimball). As professor Box once said, "all models are wrong, but some are useful." Similarly echoed by Tukey, "Far better an approximate answer to the right question, which is often vague, than an exact answer to the wrong question, which can always be made precise."

The contents of the book very much reflect the principles described above. I emphasize the importance of understanding the data before any attempt of analysis, which I learned from Mallows (Chapters 1 and 2). Since regression forms the foundation of many modeling, I include it herein (Chapter 3) with a lot more varieties which can't ever be inclusive. Even though parsimonious model derived from shrinkage is important and convenient, one needs to be careful about its purpose. For example, in genomics, gene-groups with collaborated functions often are highly correlated, to select only one or two of them can mask their importance.

Recursive partition of data (Chapter 4) has becoming ever more important in the presence of large data with heterogeneity of subject sub-populations therein. It is rarely that one can have one overall model/estimate to describe the data well, however, due to the large sample size, any hypothesis test would become significant even though the effect could be minuscule. Many analysts use this method to identify important variables and sub-populations, however, it is important to "confirm/validate" the findings with other methods for similar purposes (ref. Tukey).

Chapter 5 is about Support Vector Machine which I also had learned when I was at AT&T Bell Labs, however, it was not easy to use it for real large data in those days due to the limitation of computing power. It deals with data from a different angle by trying to identify the important support vectors instead of variables. With the current capabilities of computing and feature selection as well as the incorporation of kernels, it has become quite flexible and provide another approach to discriminate or sub-setting the data population. Graphical presentation of the results is still quite challenging.

Cluster analysis (Chapter 6) has been commonly used in genomics research even though it is not as commonly used to analyze clinical trial data. Bioinformaticians had extended this into more clustering varieties, however, the final clusters still very much depend on the metrics and linkage methods. It is prudent to conduct analysis using various linkage methods and discuss with the subject-matter experts for the proper selection and interpretation.

Neural network analysis (Chapter 7) is quite useful or controversial depending on the problems and the environments where the results are intended to be used. It can construct a network for high degree of prediction, but it can also be very difficult to interpretate except for the simple networks. It goes against the principle of parsimonious model and, therefore, how to use the model for a new data set can be quite challenging.

As many are aware that most of the data analysis, especially in regressions, the relationship between outcomes and covariates are mostly for association, and rarely can be interpreted as a causal-effect relationship. This is especially challenging for the observational studies. Professors Rubin and colleagues of Statistics, and Pearl and colleagues of Computer Science had proposed various approaches to address this topic. Chapter 8 is devoted for this topic by describing various methods including matching proposed by researchers. This has become one of the important tools in data analysis when the control group of the experiment is unobtainable due to either ethic or other reasons. Therefore, how to construct a proper well-matched control group and proper estimation of treatment effect remains as an active area of research.

I included in Chapters 9 and 10 two case studies. Chapter 9 analyzes the bankruptcies of financial institutes based on many financial performance metrics. It has a "longitudinal data flavor" even though no institution identities were available (due to privacy concern), however, with the large number of institutes included, the factors affecting bankruptcies should be consistent and informative. Chapter 10 analyzes patient treatment profile based on various treatment cycles and its relationship to the final outcomes. It is a true longitudinal data set with each subject having its own observational series. It is interesting to see how the profiles related to the final outcome, which is usually not attainable by analyze the outcome from each individual cycle.

The examples shown in the book were all programed using R language. The important codes are all included in the book; therefore, it should be quite easy to follow. I can also be reached through my email (machine4kd@gmail.com) for the questions.

This book would be impossible without the help of many of my teachers, friends, and colleagues of the various organizations, including UC San Diego, AT&T Bell Labs, and various pharmaceutical companies I had worked. I would send all of them my sincere appreciation. In addition, without the support of my family, I would not have time to do my research and finish this writing, similarly, I would like to send them my deepest gratitude.

To err is human, I am sure one can find many errors in the book and I am fully responsible for that. I will try to make my best effort to correct any errors which anyone would kindly let me know.

Kao-Tai Tsai
Berkeley Heights, New Jersey, USA
March 2021

1

Statistical Data Analysis

Statistical data analysis has a long history and, over the years, many great statisticians had discussed how it should be properly conducted. The topic has become more important nowadays due to the overwhelming speed of data generation, and how to properly extract meaningful and correct information from massive data has become ever more complicated. In the following, I briefly described the discussions from Fisher, Tukey, Huber, and Mallows of their views of this topic. It is obvious that it is impossible to be exclusive, however, these discussions will at least give us some important perspectives

1.1 Perspectives of Data Analysis

In his paper "On the Mathematical Foundations of Theoretical Statistics," Fisher [21] stated, "... the object of statistical methods is the reduction of data.... [This] is accomplished by constructing a hypothetical infinite population of which the actual data are regarded as constituting a random sample." He then identified three problems, which arise in the processes of data reduction:

1. Problems of Specification. These arise in the choice of the mathematical form of the population.

2. Problems of Estimation. These involve the choice of methods of calculating statistics, which are designed to estimate the values of the parameters of the hypothetical population.

3. Problems of Distribution. These include discussions of the distribution of statistics derived from samples.

He assumed that one can capture the essence of the real-world problem in a specification of model(s), which is assumed to be known except for some parameters to be estimated from data. His thinking had substantial influence on the development of statistics, especially mathematical statistics.

Tukey, in his monumental paper "The Future of Data Analysis" [90], stated: "For a long time I have thought I was a statistician, interested in inferences from the particular to the general. But as I have watched

DOI: 10.1201/9781003205685-1

mathematical statistics evolve, I have had cause to wonder and to doubt. And when I have pondered about why such techniques as the spectrum analysis of time series have proved so useful, it has become clear that their "dealing with fluctuations" aspects are, in many circumstances, of lesser importance than the aspects that would already have been required to deal effectively with the simpler case of very extensive data, where fluctuations would no longer be a problem. All in all, I have come to feel that my central interest is in data analysis, which I take to include, among other things: procedures for analyzing data, techniques for interpreting the results of such procedures, ways of planning the gathering of data to make its analysis easier, more precise or more accurate, and all the machinery and results of (mathematical) statistics which apply to analyzing data."

Tukey further stated, "Large parts of data analysis are inferential in the sample-to-population sense, ... Large parts of data analysis are incisive, laying bare indications which we could not perceive by simple and direct examination of the raw data, ... Some parts of data analysis, ..., are allocation, in the sense that they guide us in the distribution of effort and other valuable considerations in observation, experimentation, or analysis. Data analysis is a larger and more varied field than inference, or incisive procedures, or allocation." He further stated that data analysis must look to a very heavy emphasis on judgment. He considered at least three different sorts or sources of judgment that are likely to be involved in almost every instance: (1) judgment based upon the experience of the particular field of subject matter from which the data come, (2) judgment based upon a broad experience with how particular techniques of data analysis have worked out in a variety of fields of application, and (3) judgment based upon abstract results about the properties of particular techniques, whether obtained by mathematical proofs or empirical sampling. And it is "Far better an approximate answer to the right question, which is often vague, than an exact answer to the wrong question, which can always be made precise."

In his Fisher memorial lecture, Mallows [51], being a distinguished statistician at the AT&T Bell laboratories, who worked with scientists and engineers from various disciplines, had emphasized that there are problems that precede Fisher's, namely deciding what the relevant population is, what the relevant data are, and just how these relate to the purpose of the statistical study, in addition to choosing what problem to study.

This kind of thinking has been a required practice in many scientific industries, such as in epidemiology, clinical trials for drug discovery, etc. One often analyzes the observational data, which could be heavily confounded by unobserved factors. Then further validates the findings by conducting trials with rigorous experimental designs, which clearly specify the population of interest and the objectives of study, as well as the specifications of procedures for trial conducts.

Mallows hence integrated the thinking from Fisher and Tukey and summarized the steps of data analysis into five problems by defining the Zeroth

Problem: "Considering the relevance of the observed data, and other data that might be observed, to the substantive problem" before Fisher's specification of models and the Fourth Problem regarding the presentation of conclusions from data analysis. He also stated that "statistical thinking concerns the relation of quantitative data to a real-world problem, often in the presence of variability and uncertainty. It attempts to make precise and explicit what the data have to say about the problem of interest."

1.2 Strategies and Stages of Data Analysis

As emphasized by professor Peter Huber, the need for strategic thinking in data analysis is imposed on us by the nature of ever larger data sets. Data analysts need to consider not just the size, but also the facts that large databases are less homogeneous and have more complex internal structure. Data which is heterogeneous in precision, in variation and inevitable incomplete, all impose great challenges to analysts. Data analysis is an iterative and tedious work; therefore, it is important to focus on the important tasks and be flexible in conduct to avoid data dredging to waste time and effort

There is a broad range in the conduct of data analysis. It begins with the identification of the problem of interest, specification of the analysis protocol, and plan including the potential statistical models of interest; henceforward to plan the data collection with good data quality control.

For any extensive and complicated projects, a team of multidisciplinary subject-matter experts is very important so that the various aspects of the project can be explored to avoid the potential difficulties in execution and to produce valid analytical results.

For the planning of data collection, data analysts need to get involved if possible because this will most likely impact the data quality and make later analysis more efficient and valid. Interim data check during the collection processes is important to make sure the collection processes are followed correctly, the intended data are collected, and mistakes are corrected before a great amount of effort and resources are wasted.

For the scenario that data are already existent, it is critical for data analysts to carefully examine the meta-data if available, which usually describes the contents of the data in detail, so that the database can be better understood before planning the data analysis.

Most of the real-world projects have various assumptions for the unknown factors; therefore, it is important to conduct sensitivity analysis (or the so-called what-if analysis) after initial statistical models are fitted so that the impacts of the assumptions can be tested. It is also important to compare the results with other related findings from other research to check the validity of the findings. As professor George Box once said that no model is correct, but some are useful; therefore, flexibility in data modeling is always recommended.

To present the final results, the well thought through graphs and tables is definitely preferred to lines and lines of text. Be careful not to overly crowd the contents of the graphs and tables to the point the messages are lost, and the readers are confused. The steps described above can be tactically summarized as in Figure 1.1.

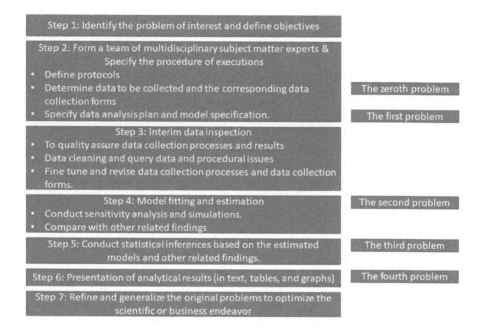

FIGURE 1.1
Data analysis strategies.

1.3   Data Quality

Data with good quality are the ingredient for analysis. There are many definitions of data quality, however, in engineering, the general scope includes

- Degree of excellence exhibited by the data in relation to the portrayal of the actual scenario.

- The state of completeness, standards based, validity, consistency, timeliness, and accuracy that make data appropriate for a specific use.

- The totality of features and characteristics of data that bear on its ability to satisfy a given purpose: the sum of the degrees of excellence for factors related to data.

- The processes and technologies involved in ensuring the conformance of data values to business requirements and acceptance criteria.

In essence, data quality is the degree to which it is complete, valid, accurate, consistent, and timely. In statistics, it adds several other complications, especially in big data.

Big data promises new levels of scientific discovery and economic value. The general expectation of big sample size is that it may give better opportunities to explore the hidden structures of each subpopulation of the data, which is traditionally not as easy when the sample size is small, and to identify important common or uncommon features across many sub populations. However, these expectations are yet to be realized in some greater degrees, primarily due to the data quality and capacity of the analysts. In the following, we describe some common challenges usually facing the analysts during the course of data analysis.

1.3.1 Heterogeneity in Data Sources

Unlike the well-designed experiments that collect well-controlled data, most of the big data are observational and coming from aggregation of several data sets. For example, some data may come from well-controlled studies, some from observational studies, and some could be from past studies, etc. This difference may reflect certain degrees of data quality heterogeneity.

1.3.1.1 Heterogeneity in Study Subject Populations

Some heterogeneity may include differences in subject populations, or regional and cultural differences, the time of data generations. For example, in clinical trials, study subjects can include various demographics such as age, gender, medical histories, genotypes, and phenotypes, etc. Even though the pharmaceutical companies try to have a more homogeneous subject population by imposing restrictive inclusion and exclusion criteria, the population homogeneity still cannot be guaranteed.

1.3.1.2 Heterogeneity in Data due to Timing of Generations

For example, due to the advance of medical research and increased biological understanding of diseases in the past decades, the medical treatment practices during later more advanced days may be very different from the treatments years ago for the same disease. These are commonly seen in the treatment of cancers, heart disease, immune system disorder, etc.; therefore, the data collected from different periods may post substantial amounts of differences in various aspects. Whether and how these kinds of data should be aggregated for analysis of treatment effects is a challenging question.

1.3.2   Noise Accumulation

Every data sample is associated with its own distribution and random error. The precision of the estimation of the parameters is affected by these random errors. These estimation errors accumulate when a model or prediction rule depends on a large number of such parameters. Researchers had suggested to alleviate the noise issue and to increase the precision involving less variables by various model assumptions and variable selections. However, variable selection in high dimensions is challenging due to spurious correlation, which is described in the next section, heterogeneity, and measurement errors, etc.

1.3.3   Spurious Correlation

It is well known that many uncorrelated random variables may have accidental high sample correlations in high dimensional data. The correlations may be real or incidental; however, given the large number of dimensions, it is very difficult to examine them sufficiently. The spurious correlation has significant impact on variable selection and may lead to false scientific discoveries. Besides variable selection, spurious correlation may also lead to wrong statistical inference. To cope with the noise accumulation issue, when the dimension d is comparable to or larger than the sample size n, it is popular to assume that only a small number of variables contribute to the response, i.e. the set of unknown parameters is a sparse vector.

1.3.4   Missing Data

Missing data are part of data collections and the problems can be more acute if data are collected over a long period of time. Even for big data, there is still the missing data problem. When $p >> n$ with moderate or small n, any missing data of the p variables from the n subjects will create the same problem as that in the moderate or small sample data. Missing data imputation may still be performed; however, it may not be as straightforward as the practices for data with smaller p due to the correlation among the variables and the issues can arise if the correlations are spurious. In addition, the large dimension and invertibility of the covariance matrix of the variables may post some additional challenges for imputations.

1.4 Data Sets Analyzed in This Book

1.4.1 NCI-60

The NCI-60 cancer cell line panel is a group of 60 human cancer cell lines used by the National Cancer Institute (NCI) for the screening of compounds to detect potential anticancer activity. The screening procedure is called the NCI-60 Human Tumor Cell Lines Screen, and it is one of the Discovery and Development Services of NCI's Developmental Therapeutics Program(DTP). Due to the diversity of the cell lines, it is possible to compare tested compounds by their effect patterns, high correlation potentially corresponding to similar effect mechanisms. The same panel is used in the Molecular Target Program for the characterization of molecular targets. Measurements include protein levels, RNA measurements, mutation status, and enzyme activity levels. The panel holds cell lines representing leukemia, melanoma, non-small-cell lung carcinoma, and cancers of the brain, ovary, breast, colon, kidney, and prostate. The NCI60 screen supported the development of several anticancer drugs, such as paclitaxel and bortezomib, which were approved by the US Food and Drug Administration (FDA) for cancer treatment (26, 27). Data mining on the NCI60 screen result also led to many findings. For example, association analysis between gene mutation status and drug efficacy on the NCI60 panel discovered that the BRAF mutation is a predictor of MEK inhibitor sensitivity.

1.4.2 Riboflavin Production with *Bacillus Subtilis*

This is a data set about riboflavin (vitamin B2) production with *B. subtilis* provided by DSM (Kaiseraugst, Switzerland) and can be found at the following website: http://www.annualreviews.org. There is a single real-valued response variable, which is the logarithm of the riboflavin production rate with $p = 4088$ covariates measure the logarithm of the expression level of 4088 genes. These gene expressions were normalized. It is a rather homogeneous data set for $n = 71$ samples that were hybridized repeatedly during a fed-batch fermentation process in which different engineered strains and strains grown under different fermentation conditions were analyzed. This data set is denoted as riboflavin.

1.4.3 TCGA

The Cancer Genome Atlas (TCGA), a landmark cancer genomics program, https://www.cancer.gov/about-nci/organization/ccg/research/structural-genomics/tcga, molecularly characterized over 20,000 primary cancer and matched normal samples spanning 33 cancer types. This joint effort between the National Cancer Institute and the National Human Genome Research Institute began in 2006, bringing together researchers from diverse disciplines and multiple institutions. Over the next dozen years, TCGA generated over 2.5 petabytes of genomic, epigenomic, transcriptomic, and proteomic data. The data, which have already lead to improvements in our ability to diagnose, treat, and prevent cancer, will remain publicly available for anyone in the research community to use.

1.4.4   The Boston Housing Data Set

A data set derived from information collected by the U. Census Service concerning housing in the area of Boston, Mass. It was obtained from the StatLib archive (http://lib.stat.cmu.edu/datasets/boston), and has been used extensively throughout the literature to benchmark algorithms. The data set is small in size with only 506 cases. The data were originally published by Harrison, D. and Rubinfeld, D.L. "Hedonic prices and the demand for clean air," J. Environ. Economics & Management, vol.5, 81–102, 1978. There are 14 attributes in each case of the data set. They are CRIM (per capita crime rate by town), ZN (proportion of residential land zoned for lots over 25,000 sq.ft.), INDUS (proportion of non-retail business acres per town), CHAS (Charles River dummy variable (1 if tract bounds river; 0 otherwise)), NOX (nitric oxides concentration (parts per 10 million)), RM (average number of rooms per dwelling), AGE (proportion of owner-occupied units built prior to 1940), DIS (weighted distances to five Boston employment centers), RAD (index of accessibility to radial highways), TAX (full-value property-tax rate per $10,000), PTRATIO (pupil-teacher ratio by town), B ($1000(Bk - 0.63)^2$, where Bk is the proportion of blacks by town), LSTAT (% lower status of the population), and MEDV (median value of owner-occupied homes in $1000s).

In the following chapters, we will present various methods which were proposed by researchers with the effort to cope with various data issues and also to produce meaningful and statistical valid results for the interpretations.

2

Examining Data Distribution

As professor Tukey once said that every data set is unique and one should understand the characteristics of the data before attempting to produce summaries or conduct statistical modeling, and the best ways to do that are via graphical methods. Professor Huber echoed this philosophy and indicated that graphical methods for exploratory data analysis are among the natural evolution of statistical data analysis (see "The evolving spiral path of statistics" in Huber (1997) [38])

In this chapter, we describe some basic tools to examine the data distributions for various dimensions. It is obvious that the following descriptions cannot be exhaustive and there are many other useful tools for special purposes that do exist in various monographs and literature. Nevertheless, this is a starting point and the graphs can all be easily produced using the many packages in R.

2.1 One Dimension

2.1.1 Histogram, Stem-and-Leaf, Density Plot

Histogram, Stem-and-Leaf, and density plot are among the most commonly used methods to examine the univariate distribution of any continuous variable. This is the first step of any data analysis, especially for statistical modeling, which assumes the data would follow a certain kind of distribution.

In histogram and density plot, one has to decide the number of categories to divide the data. Too many or too few categories will make the histogram unable to show the characteristic of the distribution. Proposals exist in the literature for the optimal bandwidth and most of the software packages had these functions built-in already, but analysts may want to define the bandwidth for special purposes of their analysis.

Stem-and-Leaf is another version of histogram with the actual data values being displayed in the plot. One advantage of Stem-and-Leaf plot is the ease of reading data values from the data. In addition, one does not need to decide the number of categories to display as in the regular histogram.

DOI: 10.1201/9781003205685-2

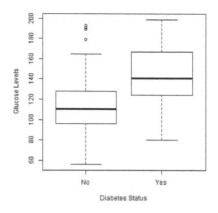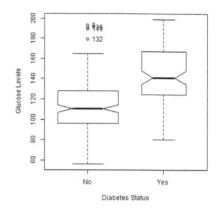

FIGURE 2.1
Box plot with variable width, notch, and w/o outliers identity.

2.1.2   Box Plot

Box plot is among one of the most used graphical methods to summarize the distribution of the univariate data. It provides five important summary statistics of the data distribution: median, lower quartile (25th percentile), upper quartile (75th percentile), and lower and upper hinges (or whiskers).

One can specify the range in the plot to determine how far the plot whiskers extend out from the box. In general, the whiskers extend to the most extreme data point, which is no more than 1.5 times the interquartile range from the box. The data outside the range of whiskers can be considered as outliers.

Options exist to make the box plot more informative. For example, for multiple box plotted together in the same graph, if the option `varwidth=TRUE`, the boxes are drawn with widths proportional to the square roots of the number of observations in the groups. Another useful option is `notch`. If `notch=TRUE`, a notch is drawn in each side of the boxes. If the notches of two plots do not overlap, which is "strong evidence" that the two medians differ as indicated by Chambers et al. [12]. The examples are shown in Figure 2.1. One can also produce a horizontal box plot by specifying `horizontal=TRUE`. For example, the following simple codes produce the box plots described above.

```
library(MASS)
library(car)
Pim<-Pima.tr2
Boxplot(glu~type, data=Pima.tr2, xlab="", ylab="")
title(xlab="Diabetes Status", ylab="Glucose Levels") #left panel
```

```
Boxplot(glu~type, data=Pima.tr2, xlab="", ylab="",
        notch=TRUE, varwidth=TRUE, id.n=10)
title(xlab="Diabetes Status", ylab="Glucose Levels") #right panel
```

2.1.3 Quantile-Quantile (Q-Q) Plot , Normal Plot, Probability-Probability (P-P) Plot

In statistical inference to compare samples, it is important to know whether the samples follow similar distributions or even have the same distributions. Q-Q plot is a useful tool that provides insight of the distributional difference by examining the whole spectrum of the distributions. In contrast, most of the statistical tests for the equality of distributions, such as the commonly used normality test or Kolmogorov-Smirnov two-sample test, only utilize the first few moments.

Q-Q plot does not require equal sample sizes between samples. In addition, many distributional aspects can be simultaneously tested. For example, shifts in location, shifts in scale, changes in symmetry, and the presence of outliers can all be detected from this plot. For example, the two data sets come from populations whose distributions differ only by a shift in location, the points should lie along a straight line that is displaced either up or down from the 45-degree reference line. If the samples are different in variances, the plot should be tilted away from the 45-degree line. The difference in tail distributions can also be detected.

More specifically, let $F(x)$ and $G(y)$ be the distributions of these two variables, and $0 < p < 1$ be the probability, then Q-Q plot is constructed by plotting

$$\{(F^{-1}(p), G^{-1}(p))\},\ 0 < p < 1. \tag{2.1}$$

Normal plot is a special case of Q-Q plot when one is interested to see whether the distribution of a data sample is normal. The normal plot can be constructed by potting

$$\{(\Phi^{-1}(p), G^{-1}(p))\},\ 0 < p < 1, \tag{2.2}$$

where Φ is the normal distribution function.

The P-P plot is another graph method which plots the percentile of one sample distribution versus the percentile of the other sample distribution. Specifically, the P-P plot can be constructed by

$$\{(p, F(G^{-1}(p)))\},\ 0 < p < 1. \tag{2.3}$$

Similarly, a normal probability plot is a special case of a P-P plot when one is interested to see whether the distribution of a variable is normal. The normal probability plot can be constructed by potting

$$\{(p, F(\Phi^{-1}(p)))\},\ 0 < p < 1. \tag{2.4}$$

Even though P-P plot is less commonly used by the practitioners; however,

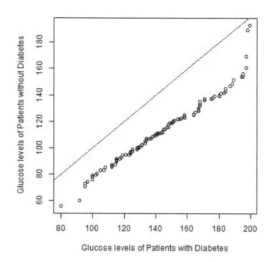

FIGURE 2.2
Q-Q plot of glucose levels for patients w/o diabetes.

the P-P plot can avoid some inferential difficulties posed by the Q-Q plot. More details can be found in Holmgren [17]. The following codes create (Figure 2.2).

```
library(MASS)
Pim<-Pima.tr2
x<-Pim[Pim$type=="Yes", 2]
y<-Pim[Pim$type=="No", 2]
qqplot(x,y, xlab="", ylab="")
abline(0,1, col="red")
title(xlab="Glucose levels of Patients with Diabetes",
      ylab="Glucose levels of Patients without Diabetes")
```

2.2   Two Dimension

2.2.1   Scatter Plot

The scatter diagram is one of the very useful tools of any data analysis. Scatter plots can show how much one variable is affected by another. A scatter plot can often be employed to identify potential associations between two variables, where one may be considered to be an explanatory variable and another may be considered a response variable. A positive association between the variables would be indicated on a scatter plot by a upward trend (positive slope), similarly, a negative association would be indicated by the opposite effect (negative slope), or, there might not be any notable association, in which case a scatter plot would not indicate any trends whatsoever.

An equation for the correlation between the variables can be estimated by some best fit procedures. For example, the linear correlation can be demonstrated by using linear regression. One of the most powerful aspects of a scatter plot, however, is its ability to show nonlinear relationships between variables. The *lowess* function (a locally weighted regression) in R is a very useful tool to identify the relationship. Furthermore, if the data are represented by a mixture model of simple relationships, these relationships will be visually evident as superimposed patterns. Even though a scatter plot can be easily produced using the command *plot(x, y)* for two variables x and y; however, the package *car* provides a collection of more comprehensive plots, which can add the smooth line and the confidence band, as well as the box plot for each variable, as shown in Figures 2.3 and 2.4.

```
library(MASS)
library(car)
scatterplot(medv ~ rm, data=Boston,
     smooth=FALSE, ellipse=FALSE, reg.line=FALSE, boxplots=FALSE,
     xlab="number of rooms", ylab="value")
title(main="Scatterplot of Boston Housing Data")

scatterplot(medv ~ rm, data=Boston,
     smooth=TRUE, ellipse=FALSE, reg.line=FALSE, boxplots=FALSE,
     xlab="number of rooms", ylab="value")
```

2.2.2   Ellipse – Visualization of Covariance and Correlation

In a linear model, a confidence region can be expressed as

$$(\beta - \hat{\beta})'X'X(\beta - \hat{\beta}) \leq s^2 pF(p, N - p, \alpha), \tag{2.5}$$

where s is the empirical variance estimated from the data and $F(p, N - p, \alpha)$ is the critical value of the F distribution with $(p, N - p)$ degrees of freedom.

In the multivariate-dimensional case, one can approximate the inference region for the parameters by

$$(\theta - \hat{\theta})'\hat{V}'\hat{V}(\theta - \hat{\theta}) \leq s^2 pF(p, N - p, \alpha). \tag{2.6}$$

For the bivariate case, the inference region can be calculated and drew according to the following relationship:

$$(x, y) = (\cos(\theta + d/2), \cos(\theta - d/2)), \tag{2.7}$$

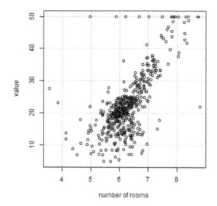

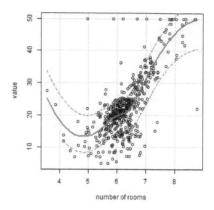

FIGURE 2.3
Simple 2-d scatter plot of two variables.

FIGURE 2.4
with smooth line and confidence band.

with $\cos(d) = \rho$, the correlation of x and y.

Using the Cars data `mtcars` of the `MASS` package in R, Figure 2.5 shows the original data plots with the ellipses of the estimated parameters about the data center and the corresponding 95% confidence regions using the estimated covariance matrix. The covariance matrix can be estimated either using the least squares method or robust methods to down-weight the possible outliers.

The shape and direction of the ellipse indicates the relative magnitude of the variances between the variables, and the positive (north-east direction) or negative (north-west direction) correlation of the estimated parameters.

Figure 2.5 shows the data ellipses, with least squares and robust estimate of covariance matrix, respectively, of the cars data that is created by the following codes. From the graph, one can easily see the positive correlation between these two variables and the magnitude of the variances.

```
library(MASS)
library(car)
data(mtcars)
dataEllipse(mtcars[,3], mtcars[,4], levels=c(0.5, 0.95), center.pch=19,
center.cex=1.5, xlab="Displacement", ylab="Horse Power")

dataEllipse(mtcars[,3], mtcars[,4], levels=c(0.5, 0.95), center.pch=19,
center.cex=1.5, robust=TRUE, xlab="Displacement", ylab="Horse Power")
```

One can also fit the variable `mpg` by the variables `disp`, `cyl`, and the covariance matrix of the estimated parameters for these variables can also

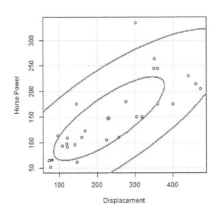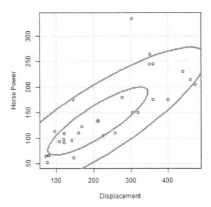

FIGURE 2.5
Least squares (left) and robust (right) estimates of data ellipse of cars data.

be graphed as shown in Figure 2.6. Similarly, one can easily see the negative correlation and the relative magnitude of the variance between these two
variables. The graph is produced by the following codes.

```
fit <- lm(mpg ~ disp + cyl , mtcars)
confidenceEllipse(fit, which.coef= c('disp', 'cyl'), levels=0.95,
Scheffe=FALSE, center.pch=19, center.cex=1.5, segments=51,
col=palette()[2], lwd=2, fill=FALSE, fill.alpha=0.3, draw=TRUE,
xlab="Displacement", ylab="Horse Power")

print("Variance-Covariance Matrix of the Estimated Parameters")
print(vcov(fit))
[1] "Variance-Covariance Matrix of the Estimated Parameters"
              (Intercept)          disp          cyl
(Intercept)   6.48722874   0.0164777387 -1.615718386
disp          0.01647774   0.0001052158 -0.006586396
cyl          -1.61571839  -0.0065863960  0.506722267
```

When the dependent variable is discrete, one can perform similar analytical
steps to get the data ellipse and the ellipse for the estimated parameters
via generalized linear modeling. We use the Pima Indian Diabetes data to
illustrate the procedures. The following sample codes produce the data ellipse
as shown in Figures 2.7.

```
library(MASS)
Pim<-Pima.tr2
```

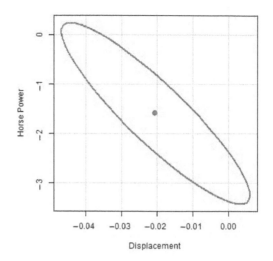

FIGURE 2.6
Covariance ellipse of the estimated parameters of cars data.

```
dataEllipse(Pim$glu, Pim$age, levels=c(0.5, 0.95), center.pch=19,
center.cex=1.5, xlab="Glucose Level", ylab="Age")

dataEllipse(Pim$glu, Pim$age, levels=c(0.5, 0.95), center.pch=19,
center.cex=1.5, robust=TRUE, xlab="Glucose Level", ylab="Age")

fit<- glm(type ~ glu + age, data=Pim,  family=binomial())
confidenceEllipse(fit, which.coef= c('glu', 'age'), levels=0.95,
Scheffe=FALSE, center.pch=19, center.cex=1.5, segments=51,
col=palette()[2], lwd=2, fill=FALSE, fill.alpha=0.3, draw=TRUE,
xlab="Glucose Level", ylab="Age")

print("Variance-Covariance Matrix of the Estimated Parameters")
print(vcov(fit))
[1] "Variance-Covariance Matrix of the Estimated Parameters"
              (Intercept)           glu             age
(Intercept)   0.608524815 -3.715419e-03 -3.200757e-03
glu          -0.003715419  3.206397e-05 -1.245096e-05
age          -0.003200757 -1.245096e-05  1.408382e-04
```

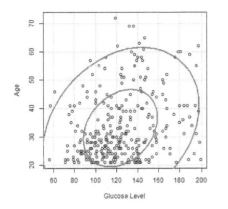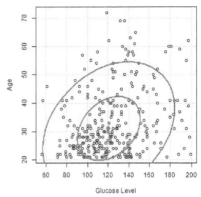

FIGURE 2.7

Least squares (left) and robust (right) estimates of data ellipse of Pima Indian diabetes data.

2.2.3 Multivariate Normality Test

Most of the joint normality tests in the literature are based on numerical calculations and most of them suffer low power to reject the null hypotheses of multivariate normality.

For graphical tests of normality, one can start with the normal plot of each marginal distribution and examine any deviation from a straight line. For multivariate responses, Andrews et al. [3] have suggested a graphical method that utilizes a *radius-and-angels* representation of multivariate data.

For a set of multivariate observations $\{\mathbf{y_i}, i = 1, \cdots, n\}$, the squared radii of the observations are defined as

$$r_i^2 = (\mathbf{y_i} - \bar{\mathbf{y}})'\mathbf{S}^{-1}(\mathbf{y_i} - \bar{\mathbf{y}}). \tag{2.8}$$

If the observations are from a p-variate normal distribution, r_i^2 should follow a χ^2 distribution with p-degrees of freedom. For example, one can use the following codes to test multivariate normality and the graphical exhibition as shown in Figure 2.8.

```
library(MASS)

Sigma <- matrix(c(10,3,3,2),2,2)
mvdata<-mvrnorm(n=1000, rep(0, 2), Sigma)
zvar<-var(mvdata)
mvmean<-apply(mvdata, 2, mean)
```

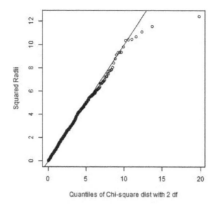 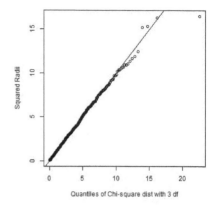

FIGURE 2.8
Bivariate and trivariate normality test.

```
radii<-diag((mvdata-mvmean)%*%solve(zvar)%*%t(mvdata-mvmean))
qchi<-qchisq(c(1:length(radii))/(length(radii)+0.05), nrow(Sigma),
ncp=0, lower.tail = TRUE, log.p = FALSE)
plot(qchi, sort(radii), xlab="", ylab="")
abline(0,1)
title(xlab="Quantiles of Chi-square dist with 2 df",
ylab="Squared Radii")

Sigma <- matrix(c(10,3,2,3,5,2,2,2,3),3,3)
mvdata<-mvrnorm(n=1000, rep(0, 3), Sigma)
zvar<-var(mvdata)
mvmean<-apply(mvdata, 2, mean)
radii<-diag((mvdata-mvmean)%*%solve(zvar)%*%t(mvdata-mvmean))
qchi<-qchisq(c(1:length(radii))/(length(radii)+0.05), nrow(Sigma),
ncp=0, lower.tail = TRUE, log.p = FALSE)
plot(qchi, sort(radii), xlab="", ylab="")
abline(0,1)
title(xlab="Quantiles of Chi-square dist with 3 df",
ylab="Squared Radii")
```

2.3 More Than Two Dimension

2.3.1 Scatter Plot Matrix

In the practice of data analysis, one usually has many variables representing the data of interest. Before any modeling or inferences, it is important to have a good understanding of the inter-relationship and its intensity between the variables, and the identification of possible outliers. Pairwise plot of each pair of variables becomes a handy tool for this purpose. A more efficient way to examine the pairwise relationship is the scatter plot matrix, which takes a few variables together to produce the plot.

Many software packages have easy ways to produce the pairwise scatter plot matrix. For example, R has the function **pairs** which produces the matrix. It also has several variations that can add more information at the discretion of the analysts. The left panel of Figure 2.9 is a simple example using Anderson's **iris** data, and the right panel of Figure 2.9 superimposes the trend line produced by the function **lowess** at the lower-left part of the plot, the histogram of each variable at the diagonal, and the pairwise correlation at the upper-right part of the plot. The R codes to produce these plots are shown below. Following codes produce examples of Scatter Plot Matrix Using pairs.

```
library(MASS)

pairs(iris[1:4], main = "Anderson's Iris Data of 3 species",
pch = 21, bg = c("red", "green", "blue")[unclass(iris$Species)])

panel.hist <- function(x, ...){
usr <- par("usr"); on.exit(par(usr))
par(usr = c(usr[1:2], 0, 1.5) )
h <- hist(x, plot = FALSE)
breaks <- h$breaks; nB <- length(breaks)
y <- h$counts; y <- y/max(y)
rect(breaks[-nB], 0, breaks[-1], y, col="cyan", ...)
}

panel.cor <- function(x, y, digits=2, prefix="", cex.cor, ...){
usr <- par("usr"); on.exit(par(usr))
par(usr = c(0, 1, 0, 1))
r <- abs(cor(x, y))
txt <- format(c(r, 0.123456789), digits=digits)[1]
txt <- paste(prefix, txt, sep="")
if(missing(cex.cor)) cex.cor <- 0.8/strwidth(txt)
text(0.5, 0.5, txt, cex = 2)
}

pairs(iris[1:4], main = "Anderson's Iris Data -- 3 species",
lower.panel=panel.smooth, diag.panel=panel.hist, upper.panel=panel.cor,
```

```
pch = 21, bg = c("red", "green", "blue")[unclass(iris$Species)])
```

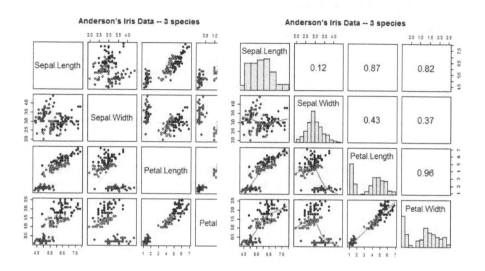

FIGURE 2.9
Pairwise scatter plot matrix.

The `scatterplotMatrix` of the `cars` package in R has several additional features, which produces other added information to the plot. By properly selecting the values of the options `diagonal, ellipse, transform, smooth, reg.line`, one can add or remove the additional information in the plot and even perform the power transformation of the variables. The R program for these plots is shown below and the corresponding plots are shown in Figures 2.10 and 2.11. The following is an example of Scatter Plot Matrix using `scatterplotMatrix`.

```
library(car)
scatterplotMatrix(iris[1:4], diagonal="density", ellipse=TRUE,
transform=TRUE, smooth=TRUE,  reg.line=FALSE, data=iris,
main="Pairwise Plot of Iris Data")
```

2.3.2   Andrews's Plot

Andrews's plot (Andrews, 1972) is a graphical technique to display high-dimensional data. The essential idea in Andrews's plot is to map each multi-dimensional observation into a function, $f(t)$, which is a linear combination of ortho-normal functions in t with the coefficients in the linear combination being the observed responses.

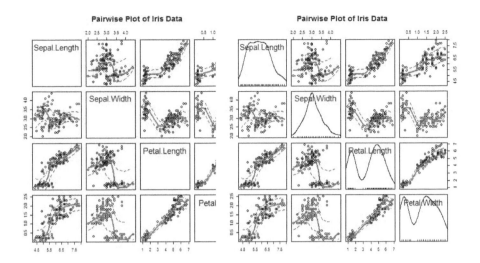

FIGURE 2.10
Pairwise scatter plot matrix.

Specifically, let

$$
\begin{aligned}
\mathbf{y_i} &= (y_{i1}, y_{i2}, \cdots, y_{ip})', \\
\mathbf{a_t} &= (a_1(t), a_2(t), \cdots)' \\
&= (1/\sqrt{2}, \sin(t), \cos(t), \sin(2t), \cos(2t), \cdots)',
\end{aligned}
$$

for $1 \leq i \leq n$ and $t \in (-\pi, \pi)$.

The n functions $\{f_1(t), f_2t, \cdots, f_n(t)\}$ can then be plotted versus t, namely,

$$\{(t, f_i(t)) \mid t \in (-\pi, \pi), i = 1, 2, \cdots, n\}. \tag{2.9}$$

Thus, the initial multi-dimensional observations will now appear as n curves in a two-dimensional display whose ordinate corresponds to the function values $f(t)$ and the abscissa is the values of t. If one changes the parameter t in the ortho-normal functions by $2\pi t$, then the plot can be created with t ranging in $(-1, 1)$.

The choice of the ortho-normal function is not unique. Tukey had suggested using

$$\mathbf{a_t} = (\cos(t), \cos(\sqrt{2}t), \cos(\sqrt{3}t), \cos(\sqrt{5}t), \cdots)' \ t \in (0, k\pi). \tag{2.10}$$

The following codes produce Andrew's plot at two different angels as shown in Figure 2.12:

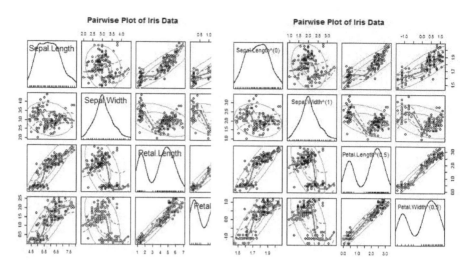

FIGURE 2.11
Pairwise scatter plot matrix.

```
library(andrews)
library(MASS)

iris2<-iris[c(1:10, 101:110),]
numstep<-100
steps<-2*pi/numstep
proj<-matrix(rep(0, numstep*nrow(iris2)), nrow(iris2), numstep)
for(j in 1:numstep){
i<-j-1
stepi<- -pi+i*steps
fa<-c(1/sqrt(2), sin(stepi), cos(stepi), sin(2*(stepi)),
cos(2*(stepi)), sin(3*(stepi)), cos(3*(stepi)))
for(kk in 1:nrow(iris2)){
proj[kk,j]<- iris2[kk, 1]*fa[1] + iris2[kk, 2]*fa[2]
+ iris2[kk, 3]*fa[3] + iris2[kk, 4]*fa[4]
}}
plot(1:numstep, proj[1,], type="n", ylim=c(min(proj)-1, max(proj)+1),
xlab="", ylab="")
title(xlab=expression(paste("Range:(", -pi, ", ", pi, ")")),
ylab="Values of Projections")
for(i in 1:10){
lines(1:numstep, proj[i,])
}
for(i in 11:nrow(proj)){
lines(1:numstep, proj[i,], col="red")
}
```

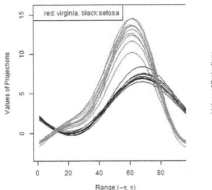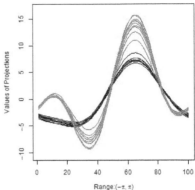

FIGURE 2.12
Andrew's plot of two different projects.

```
legend("topleft", c("red: virginia, black:setosa"))

proj<-matrix(rep(0, numstep*nrow(iris2)), nrow(iris2), numstep)
for(j in 1:numstep){
i<-j-1
stepi<- -pi+i*steps
fa<-c(1/sqrt(2), sin(stepi), cos(stepi), sin(2*(stepi)),
cos(2*(stepi)), sin(3*(stepi)), cos(3*(stepi)))
for(kk in 1:nrow(iris2)){
proj[kk,j]<- iris2[kk, 1]*fa[2] + iris2[kk, 2]*fa[3]
+ iris2[kk, 3]*fa[4] + iris2[kk, 4]*fa[1]
}}
plot(1:numstep, proj[1,], type="n", ylim=c(min(proj)-1, max(proj)+1),
xlab="", ylab="")
title(xlab=expression(paste("Range:(", -pi, ", ", pi, ")")),
ylab="Values of Projections")

for(i in 1:10){lines(1:numstep, proj[i,])}
for(i in 11:nrow(proj)){lines(1:numstep, proj[i,], col="red")}
```

2.3.3 Conditional Plot

A conditional plot, also known as a coplot or subset plot, is a plot of two variables conditional on the value of a third variable (called the conditioning variable). The conditioning variable may be either a variable that takes on only a few discrete values or a continuous variable that is divided into a

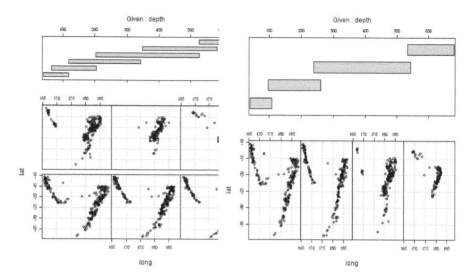

FIGURE 2.13
Conditional plots.

limited number of subsets. One limitation of the scatter plot matrix is that it cannot show interaction effects with another variable. This is the strength of the conditioning plot. It is also useful for displaying scatter plots for groups in the data. Although these groups can also be plotted on a single plot with different plot symbols, it can often be visually easier to distinguish the groups using the conditional plot.

Although the basic concept of the conditioning plot matrix is simple, there are numerous alternatives in the details of the plots. It can be helpful to overlay some type of fitted curve on the scatter plot. Although a linear or quadratic fit can be used, the most common alternative is to overlay a lowess curve. Given the variables X, Y, and Z, the condition plot is formed by dividing the values of Z into k groups. There are several ways that these groups may be formed. There may be a natural grouping of the data, the data may be divided into several equal sized groups, the grouping may be determined by clusters in the data, and so on. The page will be divided into n rows and c columns, where $n \times c \geq k$. Each row and column defines a single scatter plot. For example, the following codes create the plot (Figure 2.13) of Tonga trench earthquakes conditioning on depth of the ocean.

```
## Conditioning on 1 variable - ocean depth:
coplot(lat ~ long | depth, data = quakes)
given.depth <- co.intervals(quakes$depth, number=4, overlap=.1)
coplot(lat ~ long | depth, data = quakes, given.v=given.depth, rows=1)
```

One can also create the conditional plot by conditioning on more than one variable. The following codes produce a conditioning plot (Figure 2.14) of wool strength by conditioning on types of wool and tension strength:

```
## Conditioning on 2 factors - wool and tension:
Index <- seq(length=nrow(warpbreaks)) # to get nicer default labels
coplot(breaks ~ Index | wool * tension, data = warpbreaks,
       show.given = 0:1)
coplot(breaks ~ Index | wool * tension, data = warpbreaks,
       col = "red", bg = "pink", pch = 21,
       bar.bg = c(fac = "light blue"))
```

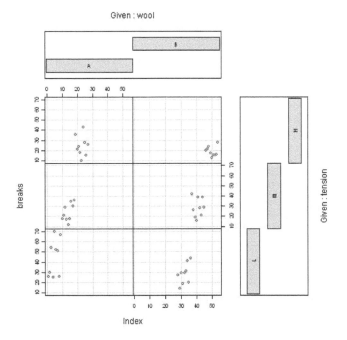

FIGURE 2.14
Example of conditional plot.

2.4 Visualization of Categorical Data

In order to explain multi-dimensional categorical data, statisticians typically look for (conditional) independence structures. Whether the task is purely exploratory or model-based, techniques such as mosaic and association plots offer good support for people to better understand the data distribution. Both aspects of (possibly higher-dimensional) contingency tables, with several extensions introduced over the last two decades, and implementations available in many statistical environments.

2.4.1   Mosaic Plot

Hartigan and Kleiner [35] proposed a mosaic plot, which is basically an area-proportional display of the observed frequencies, composed of tiles for each corresponding cells created by recursive vertical and horizontal splits of a rectangle. Thus, the area of each tile is proportional to the corresponding cell entry given the dimensions of previous splits. For example, the following simple codes produce the mosaic plot of hair x eye x sex, as shown in Figure 2.15:

```
library(MASS)
library(vcd)
data("HairEyeColor")
mosaic(HairEyeColor, shade = TRUE)
```

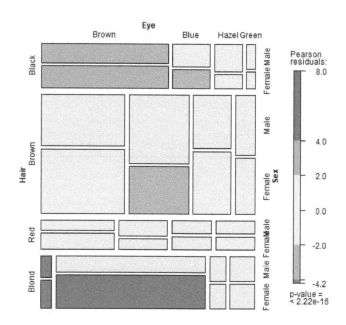

FIGURE 2.15
Mosaic plot of of hair x eye x sex.

2.4.2 Association Plot

For a contingency table, the signed contribution to Pearson's χ^2 for cell $\{ij \cdots k\}$ is

$$d_{ij\cdots k} = \frac{O_{ij\cdots k} - E_{ij\cdots k}}{\sqrt{E_{ij\cdots k}}},$$

where $O_{ij\cdots k}$ and $E_{ij\cdots k}$ are the observed and expected cell counts ,respectively, under the independence model.

Cohen [13] proposed an association plot, which displays the standardized deviations of observed frequencies from those expected under a certain independence hypothesis. In the association plot, each cell is represented by a rectangle that has (signed) height proportional to $d_{ij\cdots k}$ and width proportional to $\sqrt{E_{ij\cdots k}}$, so that the area of the box is proportional to the difference in observed and expected frequencies. The rectangles in each row are positioned relative to a baseline indicating independence $d_{ij\cdots k} = 0$. If the observed frequency of a cell is greater than the expected one, the box rises above the baseline and falls below otherwise.

Friendly [25] extended the association plot and provided a means for visualizing the residuals of an independence model for a contingency table. assoc in R package vcd is a function to produce (extended) association plots. The following codes produce two different ways to visualize multi-way table of cross-classification, as shown in Figure 2.16:

```
library(MASS)
library(vcd)
data("HairEyeColor")
(x <- margin.table(HairEyeColor, c(1, 2)))
assoc(x, main = "Relation between hair and eye color", shade = TRUE)

assoc(aperm(HairEyeColor), expected = ~ (Hair + Eye) * Sex,
    labeling_args = list(just_labels = c(Eye = "left"),
    offset_labels = c(right = -0.5), offset_varnames = c(right = 1.2),
    rot_labels = c(right = 0), tl_varnames = c(Eye = TRUE)))
```

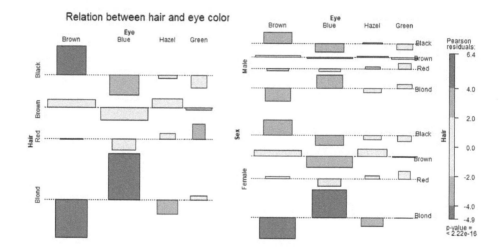

FIGURE 2.16
Two different ways to visualize multi-way table of cross-classification.

3

Regression with Shrinkage

Consider the general liner model $\mathbf{Y} = \mathbf{X}\boldsymbol{\beta} + \boldsymbol{\epsilon}$, where $\mathbf{Y} = (y_1, \cdots, y_n)^T$ is the response vector, $\mathbf{x}_j = (x_{1j}, \cdots, x_{nj})^T$, $j = 1, \cdots, p$ are the linearly independent predictors, and let $\mathbf{X} = [\mathbf{x}_1, \cdots, \mathbf{x}_p]$ be the predictor matrix, and $\boldsymbol{\beta} = (\beta_0, \beta_1, \cdots, \beta_p)$ is the unknown parameters, and $\boldsymbol{\epsilon} \sim N(0, \sigma I_{n \times n})$. Least-squares fitting procedure estimates $\boldsymbol{\beta}$ using the values that minimize

$$RSS_{\text{ls}} = \sum_{i=1}^{n}(y_i - \beta_0 - \sum_{j=1}^{p}\beta_j x_{ij})^2 = ||\mathbf{Y} - \mathbf{X}\boldsymbol{\beta}||^2. \tag{3.1}$$

Important goals of statistical learning include prediction with high accuracy and low variability, etc. via statistical models. Variable selection is particularly important with big data when $n << p$, to produce a parsimonious model, especially when the true underlying model has a sparse representation. Since least squares regression is usually not appropriate when $n << p$, alternative methods are needed. In the following, we will discuss various methods to analyze this kind of data. Generally, these methods involves regression shrinkage to derive a parsimonious model (except for Ridge regression).

3.1   Ridge Regression

Hoerl and Kennard (1988) proposed to minimize the residual sum of squares subject to a bound on the L_2-norm of the coefficients. Ridge regression is very similar to least squares, except that the coefficients are estimated by minimizing a slightly different quantity. In particular, ridge regression coefficient estimates are the values that minimize

$$RSS_{\text{ridge}} = \sum_{i=1}^{n}(y_i - \beta_0 - \sum_{j=1}^{p}\beta_j x_{ij})^2 + \lambda \sum_{j=1}^{p}\beta_j^2 \tag{3.2}$$

where $\lambda \geq 0$ is a tuning parameter, to be determined separately. (Note: Equivalently, equation (3.2) can also be written as

$$(\text{Ridge}) : \min_{\beta} \sum_{i=1}^{n}(y_i - \beta_0 - \sum_{j=1}^{p}\beta_j x_{ij})^2 \text{ subject to } \sum_{j=1}^{p}\beta_j^2 < s. \tag{3.3}$$

DOI: 10.1201/9781003205685-3

The quantity $\lambda \sum_{j=1}^{p} \beta_j^2$ is called a shrinkage penalty. When $\lambda = 0$, the penalty term has no effect, and ridge regression will produce the least squares estimates. However, as $\lambda \to \infty$, the impact of the shrinkage penalty grows, and the ridge regression coefficient estimates will approach zero.

Ridge regression's advantage over least squares is rooted in the bias-variance trade-off. As λ increases, the flexibility of the ridge regression fit decreases, leading to decreased variance but increased bias. Ridge regression does have one obvious disadvantage. Unlike best subset, forward stepwise, and backward stepwise selection, which will generally select models that involve just a subset of the variables, ridge regression will include all p predictors in the final model. The penalty $\lambda \sum \beta_j^2$ will shrink all of the coefficients toward zero, but it will not set any of them exactly to zero (unless $\lambda = \infty$).

3.2   Lasso

Tibshirani (1996) proposed lasso (Least Absolute Shrinkage and Selection Operator) by a penalized least squares method imposing an L_1-penalty on the regression. The lasso is an alternative to ridge regression that overcomes this disadvantage that ridge regression includes all p predictors in the final model. It has become very popular due to its capability of variable selection and computational feasibility. Furthermore, since the Lasso is a penalized likelihood approach, the method is rather general and can be used in a broad variety of models.

The lasso coefficients minimize the quantity

$$RSS_{\text{Lasso}} = \sum_{i=1}^{n} (y_i - \beta_0 - \sum_{j=1}^{p} \beta_j x_{ij})^2 + \lambda \sum_{j=1}^{p} |\beta_j| \tag{3.4}$$

where $\lambda \geq 0$ is a tuning parameter to be determined separately. (Note: Equivalently, equation (3.4) can also be written as

$$\min_{\beta} \sum_{i=1}^{n} (y_i - \beta_0 - \sum_{j=1}^{p} \beta_j x_{ij})^2 \text{ subject to } \sum_{j=1}^{p} |\beta_j| < s. \tag{3.5}$$

Due to the nature of the L_1-penalty, the lasso does both continuous shrinkage and variable selection simultaneously.

As mentioned previously, one of the goals of a statistical model is for prediction. To perform prediction using the Lasso is straightforward and one often uses a cross-validation (CV) scheme, e.g., 5- or 10-fold CV, to select a reasonable tuning parameter λ minimizing the cross-validated squared error risk. In addition, one can also validate the accuracy of the performance by using cross-validation.

Asymptotic results indicate that the estimates based on Lasso are consistent,

$$||(\hat{\beta}(\lambda) - \beta^0)^T(\mathbf{X}^T\mathbf{X})/n(\hat{\beta}(\lambda) - \beta^0)||_q = o_p(1) \text{ for } n \to \infty.$$

if the regularization parameter λ is of order $\sqrt{\log p/n}$ and the L_1 norms of β^0 and β_n^0 are of smaller order than $\sqrt{n/\log p}$.

Considering the accuracy for the estimated parameter $\hat{\beta}$, under compatibility assumptions [11] on the design matrix $\mathbf{X}$ and on the sparsity $s_0 = |S_0|$ in a linear model (where S_0 is the set of variables), it can be shown that

$$||\hat{\beta}(\lambda) - \beta^0||_q \to 0 \text{ in probability, if } n \to \infty$$

where $q = 1, 2$ and $||\beta||_q = (\sum_j |\beta_j|^q)^{(1/q)}$, for λ in a suitable range of order $\sqrt{\log(p)/n}$.

In practice, the tuning parameter λ is usually chosen via the cross-validation scheme aiming for prediction optimality. However, prediction optimality is often in conflict with variable selection, where the goal is to recover the underlying set of active variables S_0. For the variable selection, one often needs a larger penalty parameter than for good prediction. It is generally rather difficult to choose a proper amount of regularization for identifying the true active set S_0. When sticking to cross-validation yielding a value $\hat{\lambda}_{cv}$, the Lasso sometimes selects too many variables; therefore, by using this property, lasso can be used as a perfect tool for variable screening before a better model can be derived.

The relative performance between ridge regression and the lasso will depend on the nature of the data analyzed. Generally, one might expect the lasso to perform better in problems where a relatively small number of predictors have substantial coefficients, and the remaining predictors have coefficients that are very small, such as in many gene sequence data analyses. Ridge regression will perform better when the response is a function of many predictors, all with coefficients of roughly equal size. However, a model with many variables tends to be more difficult to interpret. In real-world data sets, one rarely knows how complicated the models will be; therefore, trying various models with proper cross-validation (preferably with new data sets) is usually recommended.

3.2.1 Example: Lasso on Continuous Data

In this section, we show the data analysis of continuous response of data set with $p >> n$. This is a data set from riboflavin production with *Bacillus subtilis* (see http://www.annualreviews.org). There is a single real-valued response variable, which is the logarithm of the riboflavin production rate. Furthermore, $p = 4,088$ (co)variables measure the logarithm of the expression level of 4,088 genes; these gene expressions were normalized with 71 rather homogeneous sample. The following are examples of the R-codes and

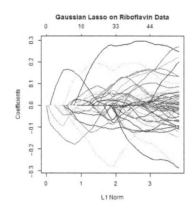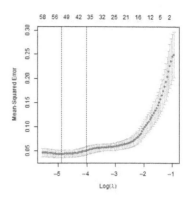

FIGURE 3.1
Model variable selection from lasso.

outputs for the analysis. The variables selected for various values of λ used and λ with the best model are shown in Figure 3.1.

```
x <- ribof3[,-1]
y <- ribof3[,1]
fit<-glmnet(x, y, family=c("gaussian"), alpha = 1, nlambda = 100)

Best lambda after CV: 0.01249693
Non-zero coefficients from minimum lambda of cross-validation:

3 4 8 232 285 412 456 581 586 600 787 965 1001 1068 1069 1109 1237 1441
1533 1585 1768 1814 1858 1914 1922 1988 2128 2261 2263 2274 3161 3193 3231
3570 3599 3614 3707 3731 3919 3961 3979 3999
```

3.2.2   Example: Lasso on Binary Data

In this section, we demonstrate the data analysis using lasso for data for binary outcomes with big sample size $(p >> n)$. This is a data set from NCI60, which has the gene expression for more than 6000 genes and 67 samples. The data set consists of 14 cancer types, which were divided into two groups of leukemia and non-leukemia type of cancers. These genes which potentially contributed to distinguish these two types of cancers are investigated using lasso. The codes for the methods are as follows, which fits the model, some selected outputs from the program and graphs are also shown in Figure 3.2.

```
x <- NCI60$data
y <- cancer2  # dichotomized cancer types
fit<-glmnet(x, y, family=c("binomial"), alpha = 1, nlambda = 100)
```

```
plot(fit, xvar = c("lambda"))
cvfit<-cv.glmnet(x, y, family=c("binomial"))
plot(cvfit)
bestlam <- cvfit$lambda.min
bestcoef<-coef(cvfit, s = cvfit$lambda.min)
print("Non-zero coefficients from minimum lambda of cross-validation")
print(t(t(bestcoef[bestcoef[,1]>0,])))

Outputs:

Best lambda = 0.0034695
Non-zero coefficients from minimum lambda of cross-validation: 2079, 2080, 2158.
```

The plot of the cross-validation and the variables entered into the model at various lasso-fitting path is shown in Figure 3.2.

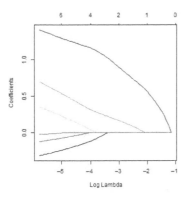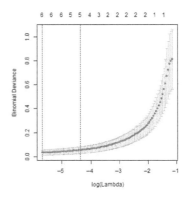

FIGURE 3.2
CV via lambdas (left panel) and the variables included in the model (right panel).

3.2.3 Example: Lasso on Survival Data

In this section, we show the data analysis of survival response of data set with $p \gg n$. This data set was from a RNA sequence data of 16,500 genes from 510 patients. Lasso is used to illustrate the data analysis. The abbreviated R codes and the outputs are shown below. The related figures are also shown in Figure 3.3.

```
xdat <- as.matrix(survival.dat)
y <- Surv(xdat$AVAL,xdat$CNSR)
fit<-glmnet(x, y, family=c("cox"), alpha = 1, nlambda = 100)
plot(fit, xvar = c("norm"))
title(main="Cox Lasso on RNA sequence Data", line=2.5)
plot(fit, xvar = c("lambda")); title(main="Cox Lasso on RNA sequence Data", line=2.5)
```

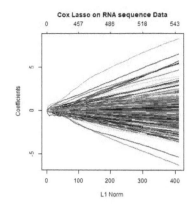 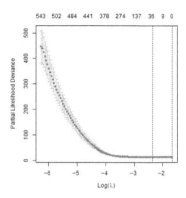

FIGURE 3.3
Model variable selection from lasso.

```
plot(fit, xvar = c("dev")); title(main="Cox Lasso on RNA sequence Data", line=2.5)
cvfit<-cv.glmnet(x, y, family=c("cox"))
plot(cvfit); title(main="Lasso on RNA sequence Data - CV of Lambda")
bestlam <- cvfit$lambda.min
print("Best lambda after CV")
print(bestlam)
bestcoef<-coef(cvfit, s = cvfit$lambda.min)
print("Non-zero coefficients from minimum lambda of cross-validation")
print(t(t(bestcoef[bestcoef[,1]!=0,])))

Outputs:
Best lambda after CV: 0.09896842
Genes with non-zero coefficients from minimum lambda of cross-validation:

HEATR6, MAP4K4, KLHL20, CPOX, CCNK, LYRM1, CCL22, KRT18, CHD4, ADGRG6, WWC1, KCNQ4, LRRC61,
MAP3K7, LMBR1L, GABPB2, CMTM7, IER2, PACSIN3, SLC20A2, TMEM42, USP32, GRM8, PSTK, PLAG1,
C9orf47, KRT8P36
```

3.3 Group Lasso

The group lasso (Yuan and Lin [102]) is a generalization of the lasso for doing group-wise variable selection. For example, in the multi-factor ANOVA problem where each factor is expressed through a set of dummy variables, deleting an irrelevant factor is equivalent to deleting a group of dummy variables. In the additive model in which each nonparametric component may be expressed as a linear combination of basis functions of the original variables, removing a component in the model is equivalent to removing a group of coefficients of the basis functions. However, the regular lasso only deletes a variable without consideration of group.

Suppose that the predictors are put into K non-overlapping groups. Consider the linear regression model, the group-lasso linear regression model solves the following penalized least squares:

$$RSS_{\text{grplasso}} = \sum_{i=1}^{n}(y_i - \beta_0 - \sum_{j=1}^{p} \beta_j x_{ij})^2 + \lambda \sum_{k=1}^{K} \sqrt{v_k} ||\beta^k||_2 \qquad (3.6)$$

where β^k is the coefficient vector of the kth group of predictors and v_k is the number of members in the kth group.

The group lasso is computationally more challenging than the lasso. Yuan and Lin implemented a blockwise descent algorithm for the group lasso penalized least squares. Meier et al. [53] developed a block-coordinate gradient descent algorithm for solving the group lasso penalized logistic regression. Meier's algorithm is implemented in a R package grplasso. In MATLAB package SLEP, Liu et al. [49] implemented Nesterov's method (Nesterov [59], [60]) for a variety of sparse learning problems. For the group-lasso case, SLEP provides functions for solving the group lasso penalized least squares and logistic regression.

Yang and Zou [100] proposed groupwise-majorization-descent algorithm to solve the general group lasso learning problems under the condition that the loss function satisfies a quadratic majorization condition. Their proposed algorithm has been implemented in a R package gglasso, which contains the functions for fitting the group lasso penalized least squares, logistic regression, Huberized SVM using the Huberized hinge loss and squared SVM using the squared hinge loss. The Huberized hinge loss and squared hinge loss are interesting loss functions for classification from machine learning viewpoint.

3.3.1 Example: Group Lasso on Gene Signatures

In this section, we illustrate the use of group lasso via R package gglasso to use three gene signatures to predict survival (above or below median survival). A number of genes in the 3 gene signatures are 2, 21, and 9, respectively. The gene signatures are denoted as groups 1, 2, and 3, as shown in the group_indices of the R program below. In the output, the final estimated coefficients of the models were based on the λ with minimum error. Four error functions, ls, logit, sqsvm, hsvm, were used to fit the models.

```
x<-as.matrix(x)        ## gene signature data
y0<-SurvTime           ## survival time
group_indices <- c(rep(1,2),rep(2,21),rep(3,9))   ## define groups

coefALL<-NULL
for(i in 1:4){
if(i==1){y<-y0}    ## regression of continuous y
if(i >1){y<-as.numeric(y0> median(y0))*2-1}    ## binary y hs to be in {-1,1}

if(i==1){fit1 <- gglasso(x=x,y=y,group=group_indices,loss="ls")}
```

```
if(i==2){fit1 <- gglasso(x=x,y=y,group=group_indices,loss="logit")}
if(i==3){fit1 <- gglasso(x=x,y=y,group=group_indices,loss="sqsvm")}
if(i==4){fit1 <- gglasso(x=x,y=y,group=group_indices,loss="hsvm")}

plot(fit1) # plots the coefficients against the log-lambda sequence
plot(fit1,group=TRUE) # plots group norm against the log-lambda sequence
plot(fit1,log.l=FALSE) # plots against the lambda sequence

if(i==1){cv <- cv.gglasso(x=x,y=y,group=group_indices,loss="ls", pred.loss="L2",
            lambda.factor=0.05, nfolds=5)}
if(i==2){cv <- cv.gglasso(x=x,y=y,group=group_indices,loss="logit", pred.loss="misclass",
            lambda.factor=0.05, nfolds=5)}
if(i==3){cv <- cv.gglasso(x=x,y=y,group=group_indices,loss="logit", pred.loss="misclass",
            lambda.factor=0.05, nfolds=5)}
if(i==4){cv <- cv.gglasso(x=x,y=y,group=group_indices,loss="logit", pred.loss="misclass",
            lambda.factor=0.05, nfolds=5)}
plot(cv)

# the coefficients at lambda.min
coefFIT<-coef(fit1,s=cv$lambda.min)
coefALL<-cbind(coefALL, coefFIT)
}

colnames(coefALL)<-c("ls","logit","sqsvm","hsvm")
print(newrownames)
rownames(coefALL)<-c("Intercept",newrownames)
print(coefALL)

## Outputs of estimates for loss functions:
```

	ls	logit	sqsvm	hsvm
Intercept	27.70760039	-2.054189344	-2.6065161144	-3.645095580
CXCL14	0.82105405	0.104809374	0.0297647608	0.051290208
CXCL17	0.42239704	0.043615114	0.0211309917	0.019040383
VSIG2	-0.07899601	0.070301657	0.0539766994	0.081712873
ST6GALNAC1	0.24457467	0.114452913	0.0846587149	0.101334121
CDH17	-0.31876580	-0.078925526	-0.0281874465	-0.036506829
CEACAM6	0.25334929	0.136492445	0.0806088827	0.133127992
LYZ	0.65233901	0.352050597	0.2163791845	0.288282364
AGR2	-0.36789282	-0.177083610	-0.1007867595	-0.136028642
ANXA10	-0.22037108	-0.032455393	-0.0330364183	-0.048671125
BTNL8	0.30691558	0.048487246	0.0213562049	0.025709491
SPINK4	0.33786444	0.069540985	0.0375982114	0.065629350
TSPAN8	-0.09790379	-0.060199121	-0.0361855045	-0.051048306
REG4	0.40903459	0.025620775	0.0066661431	0.015408134
TFF3	0.17255881	0.012215123	0.0059529672	0.010710180
TFF2	-0.39244924	-0.058677450	-0.0270575999	-0.026901128
TFF1	0.22492859	-0.024527026	-0.0177865085	-0.035444530
MYO1A	0.02215082	-0.088771010	-0.0715047757	-0.091456963
KRT20	-0.12038998	0.002080485	0.0083713169	0.006246480
LGALS4	-0.27062589	-0.160903825	-0.0865455320	-0.138991676
AGR3	-0.27659941	-0.103928661	-0.0634831299	-0.098159858
CLRN3	0.68621147	0.231710696	0.1262843326	0.166161917
CTSE	-0.06544841	-0.011927628	0.0141987217	0.045805593
FAM3D	0.07859428	0.006707702	-0.0002999041	0.004472627
TGFB1	0.00000000	-0.095990211	-0.1600879148	-0.255349933
ACTA2	0.00000000	-0.072749038	-0.0731528298	-0.119821116
SH3PXD2A	0.00000000	0.024035297	0.0443228401	0.042052356
TWIST1	0.00000000	0.009779617	0.0345415860	0.054384783
ZEB1	0.00000000	0.133114807	0.2177005757	0.317769455
TAGLN	0.00000000	-0.092677468	-0.0879138611	-0.102131926
TGFBR2	0.00000000	0.061060918	0.1054308928	0.153155126
ZEB2	0.00000000	0.078198810	0.0718700591	0.108322678
COL4A1	0.00000000	-0.018894189	0.0103479866	0.012602310

The plots of (1) the coefficients against the log-lambda sequence (Figure 3.4), (2) group norm against the log-lambda sequence (Figure 3.5), and (3) the cross-validation (Figure 3.6) are shown in below. In addition, the plots for cross-validation for error functions `logit, sqsvm, hsvm` are also shown in Figures 3.7 to 3.9 as they are somewhat different for different error function used.

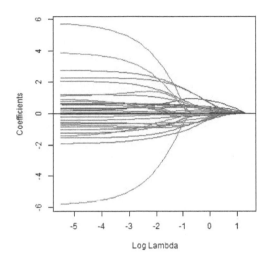

FIGURE 3.4
Group Lasso.

3.4 Sparse Group Lasso

Simon et al. [79] extended the group lasso to a more spare framework. In the high-dimensional supervised learning settings, incorporating additional problem-specific assumptions can lead to greater accuracy.

For example, with grouped covariates due to their similar biological functions, which are believed to have sparse effects on a group level, one can incorporate a more flexible regularized model for linear regression leading to group lasso.

While the group lasso gives a sparse set of groups, sometimes one would like both sparsity of groups and within each group to also identify important variables for the groups remain in the models, which leads to the model called

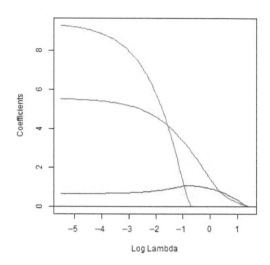

FIGURE 3.5
Group Lasso.

sparse-group lasso.

$$RSS_{sgl} = \min_{\alpha,\beta} 1/(2n)||y - X\alpha - Z\beta|| + \gamma\lambda||\beta||_1 + (1-\gamma)\lambda\sqrt{v}||\beta||_2 \quad (3.7)$$

where α and β are the coefficient vectors of the non-penalized or penalized portions of the model, respectively, with the corresponding design matrix X and Z, and the parameter γ being the mixing factor. The authors had implemented the algorithm of estimation in a R package `seagull`.

3.4.1   Example: Lasso, Group Lasso, Sparse Group Lasso on Simulated Continuous Data

In this section, we show the variable selection and estimate its effect using lasso group and sparse group lasso, and compare their precision through mean-squared errors for various of λs. The following shows the procedures using a simulation data.

```
set.seed(1000)
n <- 50 ## observations
p <- 8 ## variables

## Create a design matrix X for fixed effects to model the intercept:
```

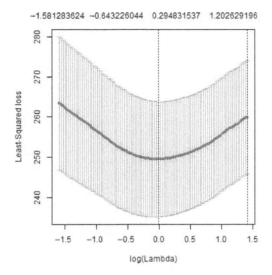

FIGURE 3.6
Group Lasso.

```
X <- matrix(1, nrow = n, ncol = 1)
## Create a design matrix Z for random effects:
Z <- matrix(rnorm(p * n, mean = 0, sd = 1), nrow = n)

## Intercept b, random effect vector u, and response y:
b <- 0.2
u <- c(0, 1.5, 0, 0.5, 0, 0, -2, 1)
y <- X%*%b + Z%*%u + rnorm(n, mean = 0, sd = 1)

## Create a vector of three groups corresponding to vector u:
group_indices <- c(1L, 2L, 2L, 2L, 1L, 1L, 3L, 1L)

## Calculate the solution:
fit_l <- seagull(y = y, X = X, Z = Z, alpha = 1.0)
fit_gl <- seagull(y = y, X = X, Z = Z, alpha = 0.0, groups = group_indices)
fit_sgl <- seagull(y = y, X = X, Z = Z, groups = group_indices)

## Combine the estimates for fixed and random effects:
estimates_l <- cbind(fit_l$fixed_effects, fit_l$random_effects)
estimates_gl <- cbind(fit_gl$fixed_effects, fit_gl$random_effects)
estimates_sgl <- cbind(fit_sgl$fixed_effects, fit_sgl$random_effects)
true_effects <- c(b, u)
print(true_effects)

## Calculate mean squared errors along the solution paths:
MSE_l <- rep(as.numeric(NA), fit_l$loops_lambda)
MSE_gl <- rep(as.numeric(NA), fit_l$loops_lambda)
MSE_sgl <- rep(as.numeric(NA), fit_l$loops_lambda)

for (i in 1:fit_l$loops_lambda) {
```

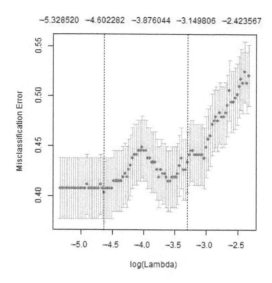

FIGURE 3.7
Group Lasso.

```
MSE_l[i] <- t(estimates_l[i,] - true_effects)%*%(estimates_l[i,] - true_effects)/(p+1)
MSE_gl[i] <- t(estimates_gl[i,] - true_effects)%*%(estimates_gl[i,] - true_effects)/(p+1)
MSE_sgl[i] <- t(estimates_sgl[i,] - true_effects)%*%(estimates_sgl[i,] - true_effects)/
(p+1)
}
minMSEindex<-order(MSE_l)[1]
fitL<-t(t(estimates_l[minMSEindex,]))
colnames(fitL)<-c("coef.min(MSEl)")

minMSEglindex<-order(MSE_gl)[1]
fitGL<-t(t(estimates_gl[minMSEglindex,]))
colnames(fitGL)<-c("coef.min(MSEgl)")

minMSEsglindex<-order(MSE_sgl)[1]
fitSGL<-t(t(estimates_sgl[minMSEsglindex,]))
colnames(fitSGL)<-c("coef.min(MSEsgl)")
print(cbind(fitL,fitGL,fitSGL))

# The estimated coefficients at the lambda with min(MSE):

     coef.min(MSEl) coef.min(MSEgl) coef.min(MSEsgl)
[1,]      0.1014887      0.14314854        0.1051350
[2,]      0.0000000      0.06827900        0.0000000
[3,]      1.3855784      1.44367805        1.3943119
[4,]      0.0000000     -0.05227492        0.0000000
[5,]      0.5707965      0.62500107        0.5820856
[6,]      0.0000000     -0.07065666        0.0000000
[7,]      0.2242504      0.32134327        0.2409049
[8,]     -2.0300066     -2.04374047       -2.0366395
[9,]      0.8251654      0.76786370        0.8226835
```

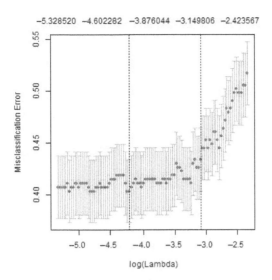

FIGURE 3.8
Group Lasso.

The columns of the output are from lasso. group lasso and sparse group lasso. The output from sparse lasso is similar to that from lasso as sparse group lasso used $\alpha = 0.9$ instead of $\alpha = 1$. One can also examine the MSE for the three approaches with various of λ using the following graph (Figure 3.10):

```
## Plot the results of the MSEs (only limited to the last 25 values for easy viewing)
plot(x = seq(1, fit_l$loops_lambda, 1)[25:50], MSE_l[25:50], type = "l", lwd = 2, xlab="",
ylab="")
abline(v=minMSEindex, lty=2)
title(main="MSE of lasso, group lasso, spare group lasso", xlab="Index of lambda values",
        ylab="Mean squared errors",
    sub="(Lines: black=lasso, blue=group-lasso, red=sparse-group-lasso)" )
points(x = seq(1, fit_l$loops_lambda, 1)[25:50], MSE_gl[25:50], type = "l", lwd = 2,
col = "blue")
abline(v=minMSEglindex, lty=2, col="blue")
points(x = seq(1, fit_l$loops_lambda, 1)[25:50], MSE_sgl[25:50], type = "l", lwd = 2,
col = "red")
abline(v=minMSEsglindex, lty=2, col="red")
```

3.4.2 Example: Lasso, Group Lasso, Sparse Group Lasso on Gene Signatures Continuous Data

In this section, we show the variable selection and estimate its effect using lasso group and sparse group lasso on the analysis of gene signature data. Three gene signatures are included. The first signature had 2 genes and will not be penalized due to its known effect in survival. The other 2 gene signatures are

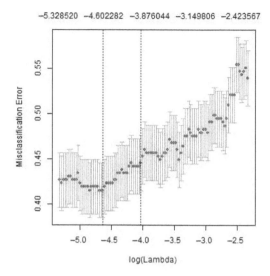

FIGURE 3.9
Group Lasso.

subject to selection of more significant genes. The estimates are shown and the MSE are compared for various of λs. The following shows the procedures using a simulation data.

```
y is the survival data
xMatrix  is the design matrix for signature 1
zMatrix  is the design matrix for signatures 2 and 3
group_indices <- c(rep(1,xdat7LEN),rep(2,xdat8LEN),rep(3,xdat9LEN))

## Calculate the solution:
fit_l <- seagull(y = y, X = xMatrix, Z = zMatrix, alpha = 1.0)
fit_gl <- seagull(y = y, X = xMatrix, Z = zMatrix, alpha = 0.0, groups = group_indices)
fit_sgl <- seagull(y = y, X = xMatrix, Z = zMatrix, groups = group_indices)

## Combine the estimates for fixed and random effects:
estimates_l <- cbind(fit_l$fixed_effects, fit_l$random_effects)
estimates_gl <- cbind(fit_gl$fixed_effects, fit_gl$random_effects)
estimates_sgl <- cbind(fit_sgl$fixed_effects, fit_sgl$random_effects)

## Calculate mean squared errors along the solution paths:
MSE_l <- rep(as.numeric(NA), fit_l$loops_lambda)
MSE_gl <- rep(as.numeric(NA), fit_l$loops_lambda)
MSE_sgl <- rep(as.numeric(NA), fit_l$loops_lambda)

p<-length(group_indices)
for (i in 1:fit_l$loops_lambda) {
MSE_l[i] <- t(estimates_l[i,])%*%(estimates_l[i,])/(p+1)
MSE_gl[i] <- t(estimates_gl[i,])%*%(estimates_gl[i,])/(p+1)
```

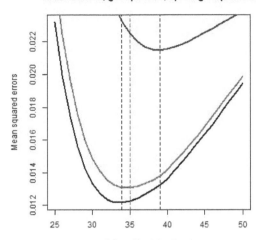

MSE of lasso, group lasso, spare group lasso

Index of lambda values
(Lines: black=lasso, blue=group-lasso, red=sparse-group-lasso)

FIGURE 3.10
MSE of lasso, group lasso, sparse group lasso.

```
MSE_sgl[i] <- t(estimates_sgl[i,])%*%(estimates_sgl[i,])/(p+1)
}
minMSEindex<-order(MSE_l)[1]
fitL<-t(t(estimates_l[minMSEindex,]))
colnames(fitL)<-c("coef.min(MSEl)")
minMSEglindex<-order(MSE_gl)[1]
fitGL<-t(t(estimates_gl[minMSEglindex,]))
colnames(fitGL)<-c("coef.min(MSEgl)")
minMSEsglindex<-order(MSE_sgl)[1]
fitSGL<-t(t(estimates_sgl[minMSEsglindex,]))
colnames(fitSGL)<-c("coef.min(MSEsgl)")
print(cbind(fitL,fitGL,fitSGL))

# The estimated coefficients at the lambda with min(MSE):

        coef.min(MSEl) coef.min(MSEgl) coef.min(MSEsgl)
 [1,]      1.31782503     1.503481374      1.57262532
 [2,]      0.52411053     0.535887452      0.53507722
 [3,]      0.00000000     0.004414696      0.00000000
 [4,]      0.00000000     0.211448846      0.00000000
 [5,]      0.00000000    -0.051915285      0.00000000
 [6,]      0.03768701     0.199319594      0.17100433
 [7,]      1.88381266     0.468714467      1.47498389
 [8,]      0.00000000    -0.009751387      0.00000000
 [9,]      0.00000000    -0.099409843      0.00000000
[10,]      0.00000000     0.178093072      0.00000000
[11,]      0.14573238     0.221262631      0.11177680
[12,]      0.00000000     0.060254179      0.00000000
[13,]      0.20436735     0.271084279      0.22068784
[14,]      0.00000000     0.126768978      0.00000000
```

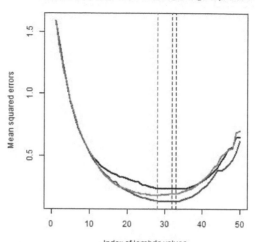

FIGURE 3.11

MSE of lasso, group lasso, sparse group lasso.

[15,]	-0.29383803	-0.221956459	-0.08844243
[16,]	0.00000000	0.122714661	0.00000000
[17,]	0.00000000	0.126493670	0.00000000
[18,]	0.00000000	-0.117808843	0.00000000
[19,]	0.00000000	-0.029618414	0.00000000
[20,]	0.00000000	-0.119895408	0.00000000
[21,]	0.60927908	0.333265685	0.45131140
[22,]	0.00000000	0.048736403	0.00000000
[23,]	0.00000000	0.073673980	0.00000000
[24,]	0.00000000	0.006586491	0.00000000
[25,]	0.00000000	0.276625462	0.00000000
[26,]	0.00000000	0.272154808	0.00000000
[27,]	0.00000000	-0.115514718	0.00000000
[28,]	1.17956598	0.437719867	0.53625189
[29,]	0.00000000	0.184589383	0.00000000
[30,]	0.00000000	0.381706563	0.18907890
[31,]	0.04477032	0.431181866	0.43355462
[32,]	0.00000000	0.295241964	0.17197027

Similar to the previous example, due to the sparseness, both lasso and sparse group lasso zero-out many variables but not in the group lasso. The first 2 genes from signature 1 were not penalized and the estimates remain similar among all three methods. One can also examine the MSE for the three approaches with various of λ using the following graph (Figure 3.11):

```
## Plot a fraction of the results of the MSEs:
plot(x = seq(1, fit_1$loops_lambda, 1), MSE_1, type = "l", lwd = 2, xlab="",ylab="",
```

```
      xlim=c(0,50), ylim=c(min(MSE_1,MSE_gl,MSE_sgl),max(MSE_1,MSE_gl,MSE_sgl)))
abline(v=minMSEindex, lty=2)
title(main="MSE of lasso, group lasso, spare group lasso", xlab="Index of lambda values",
      ylab="Mean squared errors",
      sub="(Lines: black=lasso, blue=group-lasso, red=sparse-group-lasso)" )
points(x = seq(1, fit_l$loops_lambda, 1), MSE_gl, type = "l", lwd = 2, col = "blue")
abline(v=minMSEglindex, lty=2, col="blue")
points(x = seq(1, fit_l$loops_lambda, 1), MSE_sgl, type = "l", lwd = 2, col = "red")
abline(v=minMSEsglindex, lty=2, col="red")
```

3.5 Adaptive Lasso

Assume that the true model depends only on a subset of the predictors and $A = \{j|\beta_j^0 \neq 0\}$ is the set of these predictors, and further assume that the cardinality of $|A| = p_0 < p$. Denote by $\hat{\beta}(\delta)$ the coefficient estimator produced by a fitting procedure δ. Fan and Li [18] called δ an oracle procedure if $\hat{\beta}(\delta)$ (asymptotically) has the following oracle properties:

1. Identifies the right subset model, $A = \{j|\hat{\beta}_j \neq 0\}$, and

2. Has the optimal estimation rate, $(\hat{\beta}(\delta)_A - \beta_A^*) \rightarrow_d N(0, \Sigma^*)$, where Σ^* is the covariance matrix knowing the true subset model.

Fan and Li [18] showed that the lasso can perform automatic variable selection because the L_1 penalty is singular at the origin. On the other hand, the lasso shrinkage produces biased estimates for the large coefficients, and thus it could be suboptimal in terms of estimation risk. Meinshausen and Bühlmann (2004) also showed the conflict of optimal prediction and consistent variable selection in the lasso. They proved that the optimal λ for prediction gives inconsistent variable selection results and many noise features are included in the predictive model.

Zou (2006) showed that the underlying model must satisfy a nontrivial condition if the lasso variable selection is consistent. In addition, Meinshausen and Bühlmann (2004) showed that the lasso variable selection can be consistent if the underlying model satisfies some conditions. For instance, orthogonal design guarantees the necessary condition and consistency of the lasso selection.

To fix this problem, Zou proposed the adaptive lasso, in which adaptive weights are used for penalizing different coefficients in the l_1 penalty. Adaptive lasso is a version of weighted lasso, specifically,

$$\hat{\beta}_{\text{adaptive lasso}} = \min_{\beta} |y - \sum_{i=1}^{p} x_j \beta_j|^2 + \lambda \sum_{j=1}^{p} w_j |\beta_j|,$$

with a known weight vector w. If the weights are data-dependent and cleverly chosen, then the weighted lasso can have the oracle properties. The weight w can be defined as, e.g., $w = 1/|\hat{\beta}|^{\gamma}$ for a $\gamma > 0$ and the OLS estimate of β.

In the high-dimensional situation, Buhlmann and Geer [11] suggested to use the Lasso from a first stage as the initial estimator. Typically, cross-validation is used to select the tuning parameter, e.g., $\hat{\lambda}_{\text{init,cv}}$. Thus, the initial estimator is $\hat{\beta}_{\text{init}} = \hat{\beta}(\hat{\lambda}_{\text{init,cv}})$ and let $w = 1/|\hat{\beta}_{\text{init}}|^{\gamma}$. For the second stage, one can use cross-validation as usual to select the parameter λ in the adaptive Lasso.

For the estimation of standard error, Tibshirani (1996) presented a standard error formula for the lasso. Fan and Li showed that local quadratic approximation can provide a sandwich formula for computing the covariance of the penalized estimates of the nonzero components. Zou followed the local quadratic approximation sandwich estimate to estimate the standard errors of the adaptive lasso estimates.

Zou also extended the theory and methodology to generalized linear models. Using the penalized log-likelihood estimation with adaptively weighted l_1 penalty, with the likelihood belongs to the exponential family with canonical parameter θ and density function

$$f(y|x, \theta) = h(y)\exp(y\theta - \phi(\theta)).$$

Let $\theta = x^t\beta$, and $\hat{\beta}_{mle}$ be the maximum likelihood estimates, and let the weight vector $\hat{w} = 1/|\hat{\beta}_{mle}|^{\gamma}$ for some $\gamma > 0$,

The adaptive lasso estimates $\hat{\beta}_{glm}$ are given by

$$\hat{\beta}_{glm} = \min_{\beta} \sum_{i=1}^{n}(-y_i(\mathbf{x}_i^T\beta) + \phi(\mathbf{x}_i^T\beta)) + \lambda\sum_{j=1}^{p}\hat{w}_j|\beta_j|.$$

For example, in logistic regression, the equation becomes

$$\hat{\beta}_{logistic} = \min_{\beta} \sum_{i=1}^{n}(-y_i(\mathbf{x}_i^T\beta) + \log(1 + \exp(\mathbf{x}_i^T\beta))) + \lambda\sum_{j=1}^{p}\hat{w}_j|\beta_j|.$$

Similarly, one can derive the estimates for other distributions belong to the generalized linear model family.

3.5.1   Example: Adaptive Lasso on Continuous Data

In this section, we demonstrate the data analysis using adaptive lasso using the same data in Section 3.2.1. The following are examples of the R-codes and outputs for the analysis. The variables selected for various values of λ used and the λ with the best model are shown in Figure 3.12.

```
x_bin <- ribof3[,-1]
y_bin <- ribof3[,1]

## Perform ridge regression
ridge3 <- glmnet(x = x_bin, y = y_bin, family = "gaussian", alpha = 0)
```

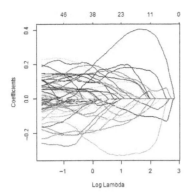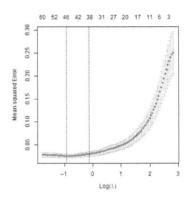

FIGURE 3.12
Model variable selection from adaptive lasso.

```
## Perform ridge regression with 10-fold CV
ridge3_cv <- cv.glmnet(x = x_bin, y = y_bin, type.measure = "deviance", nfold = 10,
                family = "gaussian", alpha = 0)
(best_ridge_coef3 <- coef(ridge3_cv, s = ridge3_cv$lambda.min))
best_ridge_coef3 <- as.numeric(best_ridge_coef3)[-1]
## Perform adaptive lasso
alasso3 <- glmnet(x = x_bin, y = y_bin, family = "gaussian", alpha = 1,
            penalty.factor = 1 / abs(best_ridge_coef3))
## Perform adaptive lasso with 10-fold cv
alasso3_cv <- cv.glmnet(x = x_bin, y = y_bin, type.measure = "deviance", nfold=10,
                family = "gaussian", alpha = 1,
                penalty.factor = 1/abs(best_ridge_coef3),keep = TRUE)
bestcoef<-coef(alasso3_cv, s = alasso3_cv$lambda.min)

Outputs:
ridge3_cv$lambda.min: 3.906184
alasso3_cv$lambda.min: 0.1706541

Non-zero coefficients from minimum lambda of cross-validation:

3 4 8 143 232 281 412 443 456 478 581 600 612 787 1109 1186 1247 1384
1441 1465 1533 1552 1585 1740 1763 1814 1858 1914 1922 1988 2035 2111 2128
2203 2261 2263 2308 2311 2743 2931 2937 2990 3193 3231 3250
3358 3540 3570 3641 3731 3806 3886 3887 39133919 3958 3961 3979 3999
```

3.5.2 Example: Adaptive Lasso on Binary Data

In this section, we demonstrate the data analysis using adaptive lasso using the same data in Section 3.2.2. The following are examples of the R-codes and outputs for the analysis. The variables selected for various values of λ used and the λ with the best model are shown in Figure 3.13.

```
x <- NCI60$data
```

```
y <- cancer2

## Perform ridge regression
ridge3 <- glmnet(x, y, family = "binomial", alpha = 0)
plot(ridge3, xvar = "lambda")

## Perform ridge regression with 10-fold CV
ridge3_cv <- cv.glmnet(x = x_bin, y = y_bin, type.measure = "deviance", nfold = 10,
                    family = "binomial", alpha = 0)
plot(ridge3_cv)

print(ridge3_cv$lambda.min)
best_ridge_coef3 <- coef(ridge3_cv, s = ridge3_cv$lambda.min)
best_ridge_coef3 <- as.numeric(best_ridge_coef3)[-1]

## Perform adaptive lasso
alasso3 <- glmnet(x = x_bin, y = y_bin, family = "binomial", alpha = 1,
penalty.factor = 1 / abs(best_ridge_coef3)) plot(alasso3, xvar = "lambda")

## Perform adaptive lasso with 10-fold cv
alasso3_cv <- cv.glmnet(x = x_bin, y = y_bin, type.measure = "deviance", nfold=10,
                    family = "binomial", alpha = 1,
                        penalty.factor = 1/abs(best_ridge_coef3),keep = TRUE)
plot(alasso3_cv)
print(alasso3_cv$lambda.min)
print("Non-zero coefficients from minimum lambda of cross-validation")
print(t(t(bestcoef[bestcoef[,1]>0,])))
[1] "Non-zero coefficients from minimum lambda of cross-validation"
38 x 5 sparse Matrix of class "dgCMatrix"
                      1           1          1           1          1
(Intercept) 0.1011349 -0.23368680 -0.24995007 0.318367950 0.06413398
128                 .           .          .           .   0.15859558
135                 .   0.09116505          .           .           .
142                 .   0.05560173          .           .           .
192                 .           .          .           .   0.28317989
231                 .   0.03531568          .           .           .
233                 .   0.12320180          .           .           .
262                 .   0.04888777          .           .           .
293                 .   0.26204952          .           .           .
311                 .           .          .  0.069540450           .
337                 .           .          .  0.194644306           .
1446                .           . 0.14143171           .           .
1867                .   0.05826869          .           .           .
2636        0.1458427           .          .           .           .
2856        0.6893243           .          .           .           .
3140                .           .          . 0.004358530           .
4079                .           . 0.01474478           .           .
4186                .           . 0.25352434           .           .
4227                .           . 0.03313285           .           .
4298                .           . 0.34544258           .           .
4644                .           . 0.19171562           .           .
4832        0.6336179           .          .           .           .
4962                .           .          . 0.109085459           .
4990        0.2276726           .          .           .           .
5504        0.5392725           .          .           .           .
5616                .           .          . 0.208268153           .
5669                .           .          . 0.033045737           .
5898                .           .          .           .   0.05881892
5899                .           .          .           .   0.37232243
6086                .           .          .           .   0.25015838
6258                .           .          .           .   0.14223159
6272                .           .          .           .   0.11644579
6410                .           .          . 0.005468184           .
6551                .           .          . 0.356912554           .
6720                .           . 0.20294992           .           .
6726                .           .          . 0.052275235           .
6765                .           .          . 0.560708986           .
```

```
6775            .         .            .       0.150713338 .
ridge3_cv$lambda.min: 3.4695
alasso3_cv$lambda.min: 0.08150069
Non-zero coefficients from minimum lambda of cross-validation: 1357, 1886, 1893, 2089, 2105,
2158
```

The plot of the cross-validation and the variables entered into the model at various lasso-fitting path is shown in Figure 3.13.

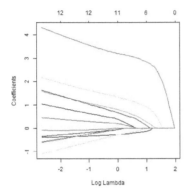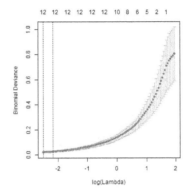

FIGURE 3.13
CV via lambdas (left panel) and the variables included in the model (right panel).

3.6   Elastic Net

As mentioned in the previous section, even though the lasso has shown success in many situations, it also has some limitations such as the following scenarios: (1) in the $p > n$ case, the lasso selects at most n variables before it saturates, (2) for a group of variables with high pairwise correlations, the lasso tends to select only one variable from the group, and (3) for usual $n > p$ situations, if there are high correlations between predictors, it has empirically observed that the prediction performance of the lasso is dominated by ridge regression.

Zou and Hastie (2005) proposed another method, namely the elastic net. Like the lasso, the elastic net simultaneously does automatic variable selection and continuous shrinkage, and it can select groups of correlated variables.

For any $\lambda_1 > 0, \lambda_2 > 0$, they defined the "naive elastic net" as

$$L(\lambda_1, \lambda_2, \beta) = |y - X\beta|^2 + \lambda_1 \sum_{j=1}^{p} |\beta_j| + \lambda_2 \sum_{j=1}^{p} \beta_j^2. \qquad (3.8)$$

The naive elastic net estimator $\hat{\beta}$ minimizes equation (3.8), namely,

$$\hat{\beta} = \min_{\beta} \{ L(\lambda_1, \lambda_2, \beta \}.$$

This procedure can be viewed as a penalized least squares method. Let $\alpha = \lambda_1/(\lambda_1 + \lambda_2)$, then equation (3.8) is equivalent to the following optimization problem

$$\hat{\beta} = \min_{\beta} |\mathbf{y} - \mathbf{X}\boldsymbol{\beta}|^2, \quad \text{subject to} \quad (1 - \alpha) \sum_{j=1}^{p} |\beta_j| + \alpha \sum_{j=1}^{p} \beta_j^2 \le t$$

for some $t > 0$. When $\alpha = 1$, it becomes a ridge regression, and lasso when $\alpha = 0$.

Computationally, the naive elastic net does the estimation using a two-stage procedure. First one finds the ridge regression coefficients, and then perform the lasso-type shrinkage along the lasso coefficient solution paths. However, the naive elastic net does not perform satisfactorily unless it is very close to ridge regression or the lasso in the regression prediction setting. To improve the performance of naive elastic net, they fine-tuned it to a better procedure, elastic net, using the following steps.

Given the data (y, X) and penalty parameters (λ_1, λ_2), and define an augmented data set $(\mathbf{y}^*, \mathbf{X}^*)$ by

$$\mathbf{X}^*_{(n+p) \times p} = \sqrt{(1+\lambda_2)} \begin{pmatrix} \mathbf{X} \\ \sqrt{\lambda_2}\mathbf{I} \end{pmatrix}, \quad \text{and} \quad \mathbf{y}^*_{(n+p)} = \begin{pmatrix} \mathbf{y} \\ 0 \end{pmatrix}.$$

The authors showed that the naive elastic net problem can be transformed into an equivalent lasso problem using the augmented data. Note that since the sample size in the augmented problem is $n+p$ and X^* has rank p, which means that the naive elastic net can potentially select all p predictors in all situations and overcomes the limitation of the lasso that were described previously. They also showed that the naive elastic net can perform an automatic variable selection in a fashion similar to the lasso. In addition, they showed that the naive elastic net has the "grouping effect" when some covariates are highly correlated.

In the attempt to correct the performance deficiency of naive elastic net, they defined the following correction of the naive elastic net and called it "elastic net" via the following approach.

Based on the augmented data, if $\hat{\beta}^*$ is the solution of the following lasso-type problem, namely,

$$\hat{\beta}^* = \min_{\beta} |y^* - X^*\beta^*|^2 + \frac{\lambda_1}{\sqrt{(1+\lambda_2)}} |\beta^*|_1,$$

then the elastic net estimator $\hat{\beta}$ can be expressed as

$$\hat{\beta}_{(\text{elastic net})} = \sqrt{1 + \lambda_2}\,\hat{\beta}^*,$$

or

$$\hat{\beta}_{(\text{elastic net})} = (1 + \lambda_2)\hat{\beta}_{(\text{naive elastic net})},$$

namely, the coefficient of the elastic net is a re-scaled coefficient of the naive elastic net. Since it is just a re-scaling, elastic net preserves all the good properties of naive elastic net.

One can further show the relationship between the elastic net estimator and the lasso estimator. Specifically, one can show that

$$\hat{\beta}_{(\text{elastic net})} = \min_{\beta} \beta^T \left(\frac{\mathbf{X}^T\mathbf{X} + \lambda_2 \mathbf{I}}{1 + \lambda_2} \right)\beta - 2\mathbf{y}^T\mathbf{X}\beta + \lambda_1 |\beta|_1, \qquad (3.9)$$

and

$$\hat{\beta}_{(\text{lasso})} = \min_{\beta} \beta^T (\mathbf{X}^T\mathbf{X}) - 2\mathbf{y}^T\mathbf{X}\beta + \lambda_1 |\beta|_1. \qquad (3.10)$$

Since

$$\frac{\mathbf{X}^T\mathbf{X} + \lambda_2 \mathbf{I}}{1 + \lambda_2} = \frac{1}{1 + \lambda_2}\mathbf{X}^T\mathbf{X} + \frac{\lambda_2}{1 + \lambda_2}\mathbf{I},$$

one can conclude that the elastic net estimator is equivalent to the lasso estimator but shrink toward the identity matrix.

3.6.1 Example: Elastic Net on Continuous Data

In this section, we demonstrate the data analysis using elastic net with the same data in Section 3.2.1. The following are examples of the R-codes and outputs for the analysis. The variables selected for various values of λ used and the λ with the best model are shown in Figure 3.14.

```
x <- ribof3[,-1]
y <- ribof3[,1]

smincv<-c(0,0,0,0)
for(ii in 1:100){
lamVal<-ii/10
## use the L1 fraction norm as the tuning parameter
fit1<-cv.enet(x, y,lambda=lamVal,s=seq(0,1,length=100),mode="fraction",
trace=FALSE,max.steps=50)
xyz<-cbind(fit1$s, fit1$cv, fit1$cv.error)
smincv<-rbind(smincv, c(lamVal,xyz[order(xyz[,2])[1],]))
}
smincv<-smincv[-1,]
smincvcv<-smincv[order(smincv[,3])[1],]
object1 <- enet(x,y,lambda=smincvcv[1], max.steps=50)
plot(object1)

Outputs:

Genes selected with estimated mu = 6.885159, lambda=0.2, and sigma2=-0.0003088
    1    2    6  214  230  241  295  305  660  912  915  932  963  999 1067 1187 1439 1562
 1582 1583 1766 1830 1920 1986 2126 2173 2216 2271 2272 2273 2464 2504 2851
 2923 3159 3194 3229 3468 3469 3470 3568 3606 3618 3798 3885 3977 4067
```

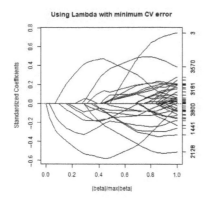
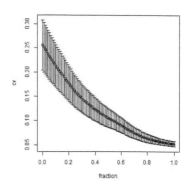

FIGURE 3.14
Model variable selection from elastic net.

3.6.2   Example: Elastic Net on Binary Data

In this section, we demonstrate the data analysis using elastic net with the same data in Section 3.2.2. The following are examples of the R-codes and outputs for the analysis. The variables selected for various values of λ used and the λ with the best model are shown in Figure 3.15.

```
x <- as.matrix(NCI60$data)
y <- cancer2
object1 <- enet(x,y,lambda=10, max.steps=50)
plot(object1)

# search for the best combination of lambda and s to minimize the cross-validation error.
smincv<-c(0,0,0,0)
for(ii in 1:100){
fit1<-cv.enet(x, y,lambda=ii/1,s=seq(0,1,length=100),mode="fraction",trace=FALSE,max.steps=50,
plot.it=FALSE)
xyz<-cbind(fit1$s, fit1$cv, fit1$cv.error)
smincv<-rbind(smincv, c(ii/10,xyz[order(xyz[,2])[1],]))
}
smincv<-smincv[-1,]
smincvcv<-smincv[order(smincv[,3])[1],]

# repeat the modeling again using the best lambda obtained from the search above.
object1 <- enet(x,y,lambda=smincvcv[1], max.steps=50)
plot(object1)

# perform the cross-validation using the best lambda obtained from the search above.
fit1<-cv.enet(x, y,lambda=smincvcv[1], s=seq(0,1,length=100),mode="fraction",
trace=FALSE,max.steps=50, plot.it=TRUE)

enet(x = x, y = y, lambda = 10, max.steps = 50)

Outputs:

Call: enet(x = x, y = y, lambda = smincvcv[1], max.steps = 50)
Var selected:  2080 2081 2079 2082 2083 6020 6019 6014 320 5928 6016 315 316 2084 6017 5997
          6018 1378 2078 5881 6009 6024 5879 5880 1860 6010 5967 2158 330 5959 6013 2108 312
```

2068 1357 5878 314 6026 6040 5887 1893 2086 6031 1936 319 303 6127 6079 5877 6008

One can find the minimum CV and the corresponding S from the fit1 output above to identify the number of variables in Figure 3.15 to be included in the model.

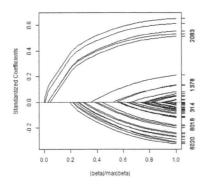 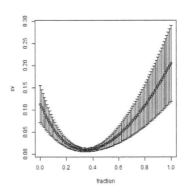

FIGURE 3.15
CV via lambdas (left panel) and the variables included in the model (right panel).

3.7 The Sure Screening Method

When the number of features p is as large as, or larger than, the number of observations n, least squares cannot (or rather, should not) be performed. The reason is simple: regardless of whether or not there truly is a relationship between the features and the response, least squares will yield a set of coefficient estimates that result in a perfect fit to the data, such that the residuals are zero. The C_p, AIC, and BIC approaches are not appropriate in the high-dimensional setting, because estimating $\hat{\sigma}^2$ is problematic. In the high-dimensional setting, the multicollinearity problem is extreme: any variable in the model can be written as a linear combination of all of the other variables in the model.

One must be careful not to overstate the results obtained and to make it clear that what one has identified is simply one of many possible models and that it must be further validated on independent data sets. When $p > n$, it is easy to obtain a useless model that has zero residuals. Therefore, one should not use the sum of squared errors, p-values, R^2 statistics, or other traditional measures of model fit on the training data as evidence of a good model fit in the high-dimensional setting. Reporting this fact might mislead others into thinking that a statistically valid and useful model has been obtained, whereas in fact this provides absolutely no evidence of a compelling model. It is important to instead report results on an independent test set, or cross-validation errors. For instance, the MSE or R^2 on an independent test set is a valid measure of model fit, but the MSE on the training set certainly is not.

One of the approaches to handle the high-dimensional data is through dimension reduction. Dimension reduction or feature selection is an important strategy to deal with high dimensionality. With dimensionality reduced from high to low, the computational burden can be reduced drastically. Meanwhile, accurate estimation can be obtained by using a well-developed lower-dimensional method.

Fan and Li suggested the concept of sure screening and proposing a sure screening method, which is based on correlation learning which filters out the features that have weak correlation with the response. The difficulties, when dimension is larger than sample size, include (1) the design matrix X which is rectangular and having more columns than rows. In this case, the matrix $X^T X$ is huge and singular. The maximum spurious correlation between a covariate and the response can be large because of the dimensionality and the fact that an unimportant predictor can be highly correlated with the response variable owing to the presence of important predictors associated with the predictor. These make variable selection difficult, (2) the population covariance matrix Σ may become ill conditioned as n grows, which adds difficulty to variable selection, (3) the minimum non zero absolute coefficient $|\beta_i|$ may decay with n and fall close to the noise level, say, the order $\{\log(p)/n\}^{-1/2}$, and (4) the distribution of dependent variable may have heavy tails. Therefore, in general, it is challenging to estimate the sparse parameter vector β accurately when $p \gg n$.

3.7.1 The Sure Screening Method

Let $M_* = \{1 \leq i \leq p : \beta_i \neq 0\}$ be the true sparse model with non-sparsity size $s = |M_*|$. The other $p - s$ variables can also be correlated with the response variable via linkage to the predictors that are contained in the model. Let $W = (w_1, \cdots, w_p)^T$ be a p-vector that is obtained by componentwise regression, i.e.

$$W = X^T y, (2)$$

where the $n \times p$ data matrix X is first standardized columnwise.

For any given $\gamma \in (0, 1)$, sort the p componentwise magnitudes of the vector w in a decreasing order and define a submodel

$$M_\gamma = \{1 \leq i \leq p : |w_i| \text{ is among the first} [\gamma n] \text{largest of all}\}, \quad (3.11)$$

where $[\gamma n]$ denotes the integer part of γn.

This is a straightforward way to shrink the full model $\{1, \cdots, p\}$ down to a submodel M_γ with size $d = [\gamma n] < n$. Such correlation learning ranks the importance of features according to their marginal correlation with the response variable and filters out those that have weak marginal correlations with the response variable. Fan & Li call this correlation screening method SIS. Thus, the idea of SIS is identical to selecting predictors by using their correlations with the response.

When there are more predictors than observations, it is well known that the least squares estimator $\hat{\beta}_{LS} = (X^T X)^+ X^T y$ is noisy, where $(X^T X)^+$ denotes the Moore-Penrose generalized inverse of $X^T X$. Consider ridge regression, let $W^\lambda = (w_1^\lambda, \cdots, w_p^\lambda)^T$ be a p-vector that is obtained by ridge regression, i.e.

$$w^\lambda = (X^T X + \lambda I_P)^{-1} X^T y$$

where $\lambda > 0$ is a regularization parameter. It is obvious that

$$w^\lambda \to \hat{\beta}_{LS} \quad as \quad \lambda \to 0, \tag{3.12}$$

and the scaled ridge regression estimator tends to the componentwise regression estimator:

$$\lambda w^\lambda \to w \text{ as } \lambda \to \infty. \tag{3.13}$$

In view of property (3.12) to make w^λ less noisy, one should choose a large regularization parameter λ to reduce the variance in the estimation. Note that the ranking of the absolute components of w^λ is the same as that of λw^λ. By equation (3.13), the componentwise regression estimator is a specific case of ridge regression with regularization parameter $\lambda = \infty$, namely, it makes the resulting estimator as little noisy as possible.

3.7.2   Sure Independence Screening on Model Selection

For the problem of ultrahigh-dimensional variable selection, we propose first to apply a sure screening method such as SIS to reduce the dimensionality from p to a relatively large scale d, say, below sample size n. Then we use a lower-dimensional model selection method such as SCAD, the Dantzig selector, lasso, or adaptive lasso. Fan & Li call SIS followed by SCAD and the Dantzig selector SIS-SCAD and SIS-DS, respectively. In some situations, one may want to reduce further the model size down to $d' < d$ by using a method such as the Dantzig selector along with hard threshold or the lasso with suitable tuning, and finally to choose a model with a more refined method such as SCAD or the adaptive lasso. Therefore, the idea of SIS makes it feasible to do model selection with ultrahigh-dimensional data.

3.7.3   Example: SIS on Continuous Data

In this section, we show the data analysis using SIS with the data as described in Section 3.2.1. The following are examples of the R-codes and outputs for the analysis.

```
x <- ribof3[,-1]
y <- ribof3[,1]
model21=SIS(x, y, family='gaussian', tune='aic')
model21=SIS(x, y, family='gaussian', tune='aic', penalty=c("lasso"))
```

```
Outputs:

SIS: tune='aic', , varISIS='vanilla', penalty=SCAD: variables selected: 1   214 2270

SIS: tune='aic', varISIS='aggr', penalty=lasso: variables selected: 295 660
SIS: tune='aic', varISIS='vanilla', penalty=lasso: variables selected: 214   295 1582 2464 3618
SIS: tune='aic', varISIS='cons', penalty=lasso: variables selected: 214   295 1878 1896 3469

SIS: tune='cv', varISIS='cons', penalty='lasso': variables selected:
        214   295   343   660 1023 1352 1851 1878 1896 2158 2634 3469 3504 3774 3984

SIS: tune='aic', varISIS='aggr', penalty=lasso: variables selected: 295 660
SIS: tune='bic', varISIS='aggr', penalty=lasso: variables selected: 295 660
SIS: tune='ebic', varISIS='aggr, penalty=lasso': variables selected: 295 660
SIS: tune='cv', varISIS='aggr', penalty=lasso: variables selected: 295 660
```

As one can easily see that the genes selected by SIS depends on very much the input parameter selection, e.g., when `varISIS=aggr`, the `tune =`. becomes less important, it always produces the same model. However, this is not as apparent with other selection of tuning parameter. In addition, it also depends on the penalizing method selected, with SCAD and lasso, they can produce very different models.

3.7.4   Example: SIS on Survival Data

In this section, we show the data analysis using SIS with the data as described in Section 3.3. The following are examples of the R-codes and outputs for the analysis.

```
xdat <- as.matrix(survival.dat)
y <- Surv(xdat$AVAL,xdat$CNSR)
model21=SIS(x, y, family='cox', penalty='lasso', tune='bic', varISIS='cons', seed=41)
model22=SIS(x, y, family='cox', penalty='lasso', tune='bic', varISIS='vanilla', seed=41)
print(model21$ix)
```

```
Outputs:

Genes selected with model21 for the model:

 SLC7A2  MAP4K4  KLHL20  DYNLL1  KRT18   CHD4   LMNB1  ECHDC2  VPS26A CDK2 SLC25A35 ZNF185
 ADCY9 KCTD5 CFL1 PLAG1 EPHB3 HNRNPAB MMP25-AS1 PSMB3

Genes selected with model22 for the model:
 HEATR6  MAP4K4  KLHL20  KRT18  CHD4  C2CD5  FBXL4  PPL  CDK2  TBC1D5  FAM149B1  GABPB2 ADAM15
 ZNF185 KCTD5 PSTK PLAG1 EPHB3 SPATS2L KPNA5
```

3.8   Identify Minimal Class of Models

In the high-dimensional setting, the task of finding the true model might be too ambitious, if meaningful at all. Even in the classical setup, with more observations than predictors, there is no model selection consistent estimator unless further assumptions are fulfilled. As professor Box once said that no model is correct but some are useful, hence instead of searching for a single "true model", Nevo and Ritov [61] proposed a number of possible models with a given number of covariates so that researchers can examine them and find potentially good prediction models. A minimal class of models could be also used in conjunction with the available models aggregation procedures, such as Bagging or Super Learner which aggregate results from ensemble of models.

The authors suggested to use a simulated annealing algorithm to search for the appropriate models via a multi-step procedure that implements both the lasso and the elastic net and then the lasso again if needed. Multi-step procedures in the high-dimensional setting have been shown to be better than the standard lasso.

Given a response variable $Y_{n \times 1}$ with covariate matrix $X_{n \times p}$ and $p > n$. Assuming $Y = X\beta^0 + \epsilon$. Denote $S_0 = \{j \mid \beta_j^0 \neq 0\}$ and s_0 be the number of elements in S_0. For any model S, define X_s to be the submatrix of X which includes only the columns specified by S. Let $\hat{\beta}^{ls}$ be the usual least squares estimator corresponding to a model S, that is,

$$\hat{\beta}_s^{ls} = (X_s^t X_s)^{-1} X_s^t Y,$$

then a straightforward approach to estimate S_0 given a model size γ is to consider the following optimization problem:

$$\min_{\beta}(1/n)\|Y - X\beta\|_2^2 \quad \text{such that} \quad \|\beta\| = \gamma.$$

The purpose of simulated annealing algorithm is to perform function optimization according to a pre-defined steps of the following objective function in model S:

$$f(S) = -(1/n)\|Y - X_s \hat{\beta}_s^{ls}\|_2^2$$

using simulated annealing with Metropolis-Hastings acceptance criterion as a search mechanism for good models. Specifically, let $S_{t,i}$ and $\hat{\beta}_{t,i}$ be the model and estimate of parameters at state (t, i), an iteration includes a suggested new model $S_{t,i}^+$, which is a minor change of the original model by exchanging/adding/deleting some variables, and obtain the corresponding new estimator $\hat{\beta}_{t,i}^+$, then a decision whether to move to this new model or stay at the original model will need to be made. This can be done by either comparing the mean-squared errors of the models or by other criteria. For example, in the classical model selection setting with $p < n$, likelihood-based criteria are often used when searching for the most appropriate model. When maximizing the likelihood to get parameters estimates becomes more complex or even infeasible as the number of possible models increases, the techniques such as simulated annealing type algorithm to implement this optimization can become useful.

3.8.1   Analysis Using Minimal Models

5-covariate Models:

```
     [,1] [,2] [,3] [,4] [,5]
[1,]   73 1820 2484 2564 4003
[2,]   73 2484 2564 3905 4003
[3,]  415  624 1425 1639 4004
[4,]   73  415  624 1639 4004
[5,]  415  624 1207 1425 4004
```

15-covariate Models:

```
     [,1] [,2] [,3] [,4] [,5] [,6] [,7] [,8] [,9] [,10] [,11] [,12] [,13] [,14] [,15]
[1,]   73 1279 1364 1503 1524 1639 1762 1827 2027  2345  2484  2564  3311  3465  4004
[2,]   73  624 1131 1516 1762 1827 1857 2027 2242  2564  3105  3226  3465  4003  4004
[3,]   73  415 1279 1425 1503 1762 1820 1855 2027  2484  2564  3905  4003  4004  4045
[4,]   73  315 1131 1210 1279 1524 1855 2564 3104  3311  3465  3514  4004  4045  4075
[5,]   73  315  415 1131 1425 1516 1524 1762 2242  2345  2564  3311  3514  4004  4075
```

41-covariate Models:

```
     [,1] [,2] [,3] [,4] [,5] [,6] [,7] [,8] [,9] [,10] [,11] [,12] [,13] [,14] [,15]
[1,]   73  415  624  827  859 1101 1131 1210 1279  1364  1425  1478  1503  1516  1524
[2,]   73  315  415  624  827  859 1101 1131 1210  1279  1364  1425  1503  1516  1524
    [,16] [,17] [,18] [,19] [,20] [,21] [,22] [,23] [,24] [,25]
[1,] 1524  1528  1636  1639  1762  1820  1827  1849  1855  1857  2027
[2,] 1528  1636  1639  1762  1820  1827  1849  1855  1857  2027
    [,26] [,27] [,28] [,29] [,30] [,31] [,32] [,33] [,34] [,35] [,36] [,37] [,38]
    [,39] [,40] [,41]
[1,] 2242  2345  2462  2484  2564  2874  3104  3105  3226  3311  3465  3514  4003
4004  4045  4075
[2,] 2242  2345  2462  2484  2564  2874  3104  3105  3226  3311  3465  3514  4003
4004  4045  4075
```

From examining the models produced by these popular methods for large data, one can realize that many models can be produced depending on the methods analysts used. There is no unique model to describe the data sufficiently, especially when the variables can be highly correlated. One important action for the analysts is to discuss the findings with subject-matter experts to understand the mechanisms of the data generations, and how the findings can be explained under this framework. It is always necessary to perform the data analysis many more times to refine the findings so that they can be implemented in the future research.

4

Recursive Partitioning Modeling

Data are the raw ingredient of statistical analysis and data homogeneity is critical for the analytical results to be interpreted with good precision and good confidence. In a well-designed experiments, specific criteria of data collection are usually defined ahead of time before data collection is taking place. For example, in clinical trials, protocols would have the subject inclusion/exclusion criteria to ensure that the targeted patients are recruited and their data collected so that the treatment effect can be proper interpreted with high degree of precision and confidence.

However, especially in large observational database, the data usually come from a variety of sub-populations. Without properly dividing the data into subpopulations accordingly, the final results can be difficult to interpret and usually with poor precision. Some usual attempts to divide subject populations by one or two variables, such as gender and age, can be grossly insufficient. Sub-populations can be different due to multiple characteristics, some of them are even unknown and not collected in database. Therefore, some kinds of recursive partitioning of the database by possibly multiple variables to properly identify the sub-populations or subgroups are crucial and should be in the toolbox of data analysts.

The monograph "Classification and Regression Tree" by Brieman et al. [9] discussed the tree-based model. They had detailed discussion of the construction of tree and the applications to data analysis. Several extensions in methodologies and computer programs for implementations had since been developed for the tree-based analysis.

4.1 Recursive Partitioning Modeling via Trees

4.1.1 Elements of Growing a Tree

To construct a tree and subsequently growing a tree, one needs to consider the following basic criteria:

1. A set of binary questions to guide the branch splits for binary trees.

2. A split criterion that can be evaluated for any split of any node

DOI: 10.1201/9781003205685-4

based on the purification of the information and improvement of interpretability of the nodes.

3. A stop-splitting rule to avoid overly splitting the trees (as over-fitting in regression).

4. A rule for assigning every terminal node to a class.

Starting from the root node, the tree algorithm searches through the variables one by one, from x_1 to x_M, if there are M variables in the data. For each variable it finds the best split. Then it compares the M best single variable splits and select the best among the best. It is feasible to perform this split since there are a finite number of variables and values for each variable. This procedure is equally applicable regardless whether the data are continuous or ordered categorical. One can apply the splits to construct a tree structure classifiers by repeating splits of subsets of data space into descendant subsets, beginning from the root note until it reaches the stop-splitting rules.

4.1.1.1   Grow a Tree

To grow a tree, one selects from the root note T_1. At the root node T_1, a search is made through all candidate splits of all candidate covariate to find that split at value s^* which gives the largest decrease in impurity, which will be defined later, namely,

$$\Delta I(s^*, T_1) = \max_{s \in S} \Delta I(s, T_1), \tag{4.1}$$

where $\Delta I(s, T_1)$ is the change of impurity by splitting at value s at current node T_1. Then T_1 is split into two "daughter-nodes" T_2 and T_3 using the split at value s^* of the best selected covariate. The same search procedure for the best value s of all candidate covariates is repeated on both T_2 and T_3 separately to grow the next generation of "daughter-nodes." When a node T is reached such that no significant decrease in impurity is possible, then T is not split further and becomes a terminal node.

Let $p(j|T)$ be the probability for an observation to belong to category j at note T, then the class character of a terminal node is determined by the following rule: if

$$p(j_0|T) = \max_j p(j|T), \tag{4.2}$$

then T is designated as a class j_0 terminal node.

4.1.2   The Impurity Function

Generally, in the regression settings, a better model is one that reduces more residual sum of squares after the model is fitted. Similar to this concept, a better tree structure classifier is one that would reduce the "impurity" of the entire tree as defined by Brieman et al.

4.1.2.1 Definition of Impurity Function

The impurity function of a node t with J possible categories is defined as

$$i(t) = \Phi\Big(p(1|t), p(2|t), \cdots, p(J|t)\Big), \tag{4.3}$$

where the impurity function, Φ, is defined on the set $S = \{(p_1, p_2, \cdots, p_J) \mid p_j \geq 0, \; \sum_j p_j = 1\}$ with the following properties:

1. $\max \Phi(\cdot) = \Phi(1/J, \cdots, 1/J)$.
2. $\min \Phi(\cdot) = \Phi(1, 0, \cdots, 0) = \cdots = \Phi(0, \cdots, 0, 1)$.
3. Φ is a symmetric function of $(p_1, p_2, \cdots, p_J)$.

When node t is split into t_l (left-daughter-node) and t_r (right-daughter-node) with a proportion p_l of the cases in t go to t_l and a proportion p_r go into t_r, the goodness of split at node t is defined as the decrease in impurity

$$\Delta i(s, t) = i(t) - p_l i(t_l) - p_r i(t_r), \tag{4.4}$$

and the impurity of the whole tree is defined as the total impurity from all of terminal notes.

4.1.2.2 Measure of Node Impurity – the Gini Index

Given a node t with estimated class probability of each node: $\{p(j|t) \mid j = 1, \cdots, J\}$, then a measure of node impurity given node t is defined as

$$i(t) = \Phi\Big(p(1|t), \cdots, p(J|t)\Big) = \sum_{i \neq j} p(j|t)p(i|t) = 1 - \sum_j p^2(j|t), \tag{4.5}$$

which is called Gini index.

Gini index can be interpreted as the empirical variance of classification (Light and Margolin [47]). For example, at node t, if one assigns all class j objects value 1 and assigns objects of all other classes value 0, then the sample variance of these values is $p(j|t)(1 - p(j|t))$. After repeating this process for all classes and sum up the variances, the result will be identical to the index defined in equation (4.5).

4.1.3 Misclassification Cost

The purpose of a recursive partitioning is to produce a structure with more homogeneous nodes so that one can maximize the probability of correct classification of subjects in a node, for example, to correctly classify the disease status of patients so that they can be properly treated with correct medicines. However, incorrect classifications are inevitable and will carry a corresponding misclassification cost.

The misclassification cost of a class j subject as class i subject, $C(i|j)$, is a function such that

$$C(i|j) = \begin{cases} > 0 & \text{if } i \neq j, \\ = 0 & \text{if } i = j. \end{cases} \quad (4.6)$$

Given a node t with estimated node probabilities $p(j|t), j = 1, \cdots, J$, if a randomly selected object of unknown class falls into t and is classified as class i, then the estimated misclassification cost at node t is $\sum_j C(i|j)p(j|t)$.

One can also define the misclassification cost at the level of a classifier. Specifically, for a classifier $\mathcal{C}$, let

$$R_j = \sum_i C(i|j)p(\mathcal{C}(x) = i \mid x = j)$$

be the expected cost of classifying a class j item, where $p(\mathcal{C}(x) = i \mid x = j)$ is the probability of classifying a class j subject as class i, then the misclassification cost for classifier $\mathcal{C}$ can be defined as

$$R(\mathcal{C}) = \sum_j R_j \pi(j)$$

where $\pi(j)$ is the probability of class j in the population.

The misclassification cost can be estimated using either an independent test sample if that is available, or through cross-validation. The use of independent sample is preferred when the learning sample contains a large number of cases. Cross-validation is computationally expensive but it makes more effective use of all cases and give useful information regarding to stability of the tree structure. Unless the sample size is quite large; otherwise, the cross-validation method is the preferred estimation method.

4.1.4   Size of Trees

If a tree is too large with many branches and nodes, similar to over-fitting a statistical model, the misclassification error tends to be large. On the other hand, if a tree is too small with few branches and nodes, similar to under-fitting a statistical model, the misclassification error can be too large due to under utilization of available information contained in the data set. Therefore, to strike a balance of creating a right-sized tree is important to properly classify the data.

A general strategy is to create a large tree initially and gradually prune upward by considering the cost of complexity of a tree. Specifically, for a tree T, let $|T|$ be the number of terminal nodes and $R(T)$ be the corresponding cost of misclassification, define the cost-complexity measure $R_\lambda(T)$ as

$$R_\lambda(T) = R(T) + \lambda|T| \quad (4.7)$$

with λ being the tuning parameter. The idea is very similar to the modeling using a penalized regression, which minimizes the mean-squared errors with a penalized factor. Then the smallest subtree $T(\lambda)$ for a complexity parameter λ is defined as a tree, which satisfies the following conditions:

$$R_\lambda(T(\lambda)) = \min_{\{T \subset T_{\max}\}} R_\lambda(T)$$

and

$$\text{if } R_\lambda(T) = R_\lambda(T(\lambda)), \text{ then } |T(\lambda)| < |T|.$$

The value of tuning parameter λ can belong to a large interval of real numbers, but since there can only be a finite number of sub-trees, the number of sub-trees is usually a step function of the values of λ, which can be presented in a graph of λ against the size of a tree.

The other useful idea is to use the values of C_p, which is usually part of the summary output, to control the tree size. The example of using C_p is shown in the examples below.

4.1.5 Example of Recursive Partitioning

In this section, we show how recursive partitioning can be implemented with the R-package **rpart** to analyze binary, continuous, and survival outcome data from TCGA pancreatic cancer and Riboflavin (vitamin B2) production with *B. subtilis*. These data have genomics data, which will be used as covariates, and the outcomes, which will be used as outcomes. Details of the data sets were described in Chapter 1.

4.1.5.1 Recursive Partitioning with Binary Outcomes

The baseline tumor grades were categorized into two groups, the grade level ≤ 2, the less severe patients, and grade levels ≥ 3, more severe patients. The **rpart** was applied to this binary outcome data to investigate the genomics effect on tumor grades.

The following shows the R-codes to analyze the data and draw the tree structure. It shows the creation of binary partition values for the variables used. The whole tree is quite large, in the following, only a smaller tree is shown using $C_p = 0.05$ as an example.

```
library(rpart)
library(rpart.plot)

y<-factor(TCGA4Tree$GradeN)  # the tumor grade data
x<-as.matrix(TCGA4Tree[,-c(1:6)])  # the gene sequence data
rpartTCGA<-rpart(y ~ x, control = rpart.control(cp = 0.05))
print(summary(rpartTCGA))

Call: rpart(formula = y ~ x, control = rpart.control(cp = 0.05))
```

n= 177

#CP information (one can use this to control the tree size):
```
          CP nsplit rel error   xerror      xstd
1 0.21153846      0 1.0000000 1.000000 0.1165378
2 0.11538462      1 0.7884615 1.134615 0.1206083
3 0.09615385      2 0.6730769 1.365385 0.1253984
4 0.05000000      5 0.3846154 1.384615 0.1256814
```

Variable importance
ARHGAP23   SLC9A8   TFAP2B   SLC4A2   TAF5L   FMO5   ...

#History of variable partitioning (only the beginning and end are shown here):
Node number 1: 177 observations,   complexity param=0.2115385
  predicted class=1 expected loss=0.2937853 P(node) =1
    class counts: 52   125
   probabilities: 0.294 0.706
  left son=2 (23 obs) right son=3 (154 obs)
  Primary splits:
    xARHGAP23 < 1.0281   to the right, improve=10.485850, (0 missing)
    xTMEM40   < 0.8122   to the right, improve= 9.988752, (0 missing)
    xHIST1H3I < 0.2249   to the right, improve= 9.733196, (0 missing)
    xIDH3A   < -0.67935 to the left, improve= 8.700277, (0 missing)
    xFGF13   < -0.6641   to the left, improve= 8.399414, (0 missing)
  Surrogate splits:
    xPTGES   < 1.0744   to the right, agree=0.921, adj=0.391, (0 split)
    xFAM134B < -1.1385   to the left, agree=0.915, adj=0.348, (0 split)
    xPYGL   < 1.9802   to the right, agree=0.915, adj=0.348, (0 split)
    xSNCG   < 0.98665   to the right, agree=0.915, adj=0.348, (0 split)
    xTMEM40   < 0.90925   to the right, agree=0.915, adj=0.348, (0 split)
...

Node number 14: 31 observations,   complexity param=0.09615385
  predicted class=1 expected loss=0.4516129 P(node) =0.1751412
    class counts: 14   17
   probabilities: 0.452 0.548
  left son=28 (16 obs) right son=29 (15 obs)
  Primary splits:
    xSLC9A8   < 0.11805   to the right, improve=8.613172, (0 missing)
    xPMEL   < -0.4299   to the left, improve=7.627566, (0 missing)
    xLGALS17A < -0.15755 to the right, improve=7.136657, (0 missing)
    xCAMKV   < -0.11735 to the right, improve=6.485273, (0 missing)
    xSLC9A3   < 0.06265   to the left, improve=6.286463, (0 missing)
  Surrogate splits:
    xTAF5L   < 0.70235   to the left, agree=0.871, adj=0.733, (0 split)
    xACCS   < -0.09635 to the right, agree=0.839, adj=0.667, (0 split)
    xCNOT2   < -0.08535 to the left, agree=0.839, adj=0.667, (0 split)

```
      xFBX028 < 0.7272    to the left,  agree=0.839, adj=0.667, (0 split)
      xGL01   < -0.6345  to the left,  agree=0.839, adj=0.667, (0 split)

Node number 15: 107 observations
  predicted class=1  expected loss=0.09345794  P(node) =0.6045198
     class counts:     10    97
   probabilities: 0.093 0.907


Node number 28: 16 observations
  predicted class=0  expected loss=0.1875  P(node) =0.09039548
     class counts:     13     3
   probabilities: 0.812 0.188


Node number 29: 15 observations
  predicted class=1  expected loss=0.06666667  P(node) =0.08474576
     class counts:      1    14
   probabilities: 0.067 0.933


n= 177


#Summary of the tree:
node), split, n, loss, yval, (yprob)
      * denotes terminal node

 1) root 177 52 1 (0.29378531 0.70621469)
   2) xARHGAP23>=1.0281 23   6 0 (0.73913043 0.26086957)
     4) xSLC4A2< 0.5936 16   0 0 (1.00000000 0.00000000) *
     5) xSLC4A2>=0.5936 7   1 1 (0.14285714 0.85714286) *
   3) xARHGAP23< 1.0281 154 35 1 (0.22727273 0.77272727)
     6) xTFAP2B>=-0.10355 16   5 0 (0.68750000 0.31250000) *
     7) xTFAP2B< -0.10355 138 24 1 (0.17391304 0.82608696)
      14) xFM05>=0.66335 31 14 1 (0.45161290 0.54838710)
        28) xSLC9A8>=0.11805 16   3 0 (0.81250000 0.18750000) *
        29) xSLC9A8< 0.11805 15   1 1 (0.06666667 0.93333333) *
      15) xFM05< 0.66335 107 10 1 (0.09345794 0.90654206) *
```

One can also use the R function **rpart.plot(rpartTCGA)** to show the recursive partitions graphically (Figure 4.1). As a general practice, a large tree is difficult for users to grasp the essence even though it shows lots of details. One can prune the tree to a proper size using the value of **Cp** for that purpose. There are other methods to control the tree size in the R package manual.

4.1.5.2 Recursive Partitioning with Continuous Outcomes

The **Riboflavin production with Bacillus subtilis** data, as described in Chapter 1, were used for this example. This is a data set about riboflavin (vitamin B2) production with *B. subtilis*. There is a single real-valued response variable, which is the logarithm of the riboflavin production rate with $p = 4088$ covariates measure the logarithm of the expression level of 4088 genes.

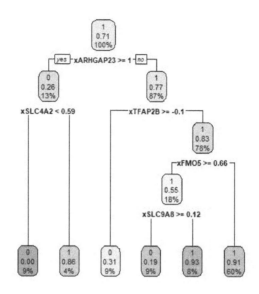

FIGURE 4.1
Graphical exhibition of recursive partitioning.

```
library(rpart)
library(rpart.plot)

Gene<-riboflavin[-1,1]
ribof1<-riboflavin[,-1]; ribof2<-ribof1[-1,]
ribof3<-data.frame(t(ribof2)); colnames(ribof3)<-Gene
x <- ribof3[,-1]
y <- ribof3[,1]
rpartRIBO<-rpart(y ~ ., data=ribof3)
print(summary(rpartRIBO))

Call: rpart(formula = y ~ ., data = ribof3)
n= 71
          CP nsplit rel error    xerror      xstd
1 0.67044752      0 1.0000000 1.0209755 0.20101135
2 0.15168969      1 0.3295525 0.3839174 0.06259746
3 0.02006192      2 0.1778628 0.2153899 0.05978176
4 0.01604922      3 0.1578009 0.1904623 0.06019719
5 0.01000000      4 0.1417517 0.1780188 0.06037737

Variable importance
```

```
AADK_at YBFE_at DHBE_at FLHB_at  GDH_at  MAF_at YKRS_at YSAA_at ...
```

```
# History of recursive partitions:
Node number 1: 71 observations,      complexity param=0.6704475
   mean=6.885159, MSE=0.2475816
Node number 2: 54 observations,      complexity param=0.1516897
   mean=6.656562, MSE=0.06597897
Node number 3: 17 observations
   mean=7.611288, MSE=0.1311828
Node number 4: 27 observations,      complexity param=0.02006192
   mean=6.434349, MSE=0.01872102
Node number 5: 27 observations,      complexity param=0.01604922
   mean=6.878775, MSE=0.01447965
Node number 8: 11 observations
   mean=6.296515, MSE=0.006505725
Node number 9: 16 observations
   mean=6.52911, MSE=0.005078141
Node number 10: 18 observations
   mean=6.806495, MSE=0.005282312
Node number 11: 9 observations
   mean=7.023335, MSE=0.001527886


n= 71
node), split, n, deviance, yval
      * denotes terminal node

 1) root 71 17.57830000 6.885159
   2) AADK_at< 7.130165 54   3.56286400 6.656562
     4) AADK_at< 6.658304 27   0.50546750 6.434349
       8) AADK_at< 6.417094 11   0.07156298 6.296515 *
       9) AADK_at>=6.417094 16   0.08125026 6.529110 *
     5) AADK_at>=6.658304 27   0.39095050 6.878775
      10) AADK_at< 6.936089 18   0.09508161 6.806495 *
      11) AADK_at>=6.936089 9   0.01375097 7.023335 *
   3) AADK_at>=7.130165 17   2.23010700 7.611288 *
```

One can also use the R function **rpart.plot(rpartRIBO)** to show the recursive partitions graphically, as shown in Figure 4.2.

4.1.5.3   Recursive Partitioning for Survival Outcomes

In this section, we analyze the overall survival data from a pancreatic cancer data set in TCGA. The survival and the censoring status are from the clinical data. The genomics data were used as the covariates. The following shows the R-codes to analyze the data and draw the tree structure.

```
library(rpart)
library(rpart.plot)
```

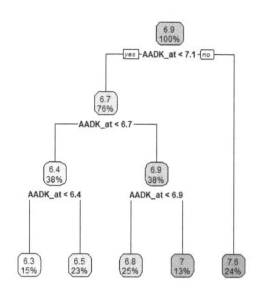

FIGURE 4.2
Graphical exhibition of recursive partitioning.

```
# survival & gene data
xx<-cbind(Clin4RNAseq$AVAL, Clin4RNAseq$CNSR,
RNAseqSubjIDn[RNAallPsig2[,2]])
xx<-data.frame(xx)  # input data has to be a data.frame

# the size of tree is quite large, use cp=0.02 to prune the tree size.
rpart.obj <- rpart(Surv(Clin4RNAseq.AVAL, Clin4RNAseq.CNSR) ~., data=xx,
control = rpart.control(cp = 0.02))
print(summary(rpart.obj))

Call: rpart(formula = Surv(Clin4RNAseq.AVAL, Clin4RNAseq.CNSR) ~ .,
    data = xx, control = rpart.control(cp = 0.02))
n= 515

          CP nsplit rel error   xerror       xstd
1 0.03975671      0 1.0000000 1.003897 0.04600623
2 0.03460949      2 0.9204866 1.018775 0.04962869
3 0.02656726      3 0.8858771 1.035515 0.05276740
4 0.02217067      4 0.8593098 1.077724 0.05880120
5 0.02183946      5 0.8371391 1.094962 0.05963086
6 0.02163511      6 0.8152997 1.094962 0.05963086
```

```
7 0.02073444       8 0.7720295 1.122940 0.06117914
8 0.02000000       9 0.7512950 1.155798 0.06478557
```

```
# important variables
FCHSD2  FAM122A  RUVBL2  TLR1  RBBP6  SRSF9 ...
```

```
#History of variable partitioning (only the beginning and end are shown
here):
Node number 1: 515 observations,    complexity param=0.03975671
  events=265,  estimated rate=1 , mean deviance=1.344074
  left son=2 (507 obs) right son=3 (8 obs)
  Primary splits:
    FCHSD2   < 5.794435    to the right, improve=34.72986, (0 missing)
    AHNAK2   < 11.24015    to the left,  improve=31.30361, (0 missing)
    HMGA1    < 7.82757     to the left,  improve=30.66661, (0 missing)
    PROSER2  < 2.358689    to the left,  improve=30.02789, (0 missing)
    LDHB     < 7.661128    to the left,  improve=29.17415, (0 missing)
  Surrogate splits:
    PTGDR    < -1.483115   to the right, agree=0.990, adj=0.375, (0 split)
    YBX3     < 7.824068    to the left,  agree=0.988, adj=0.250, (0 split)
    PHACTR3  < 5.48759     to the left,  agree=0.988, adj=0.250, (0 split)
    KPNB1    < 8.948445    to the left,  agree=0.988, adj=0.250, (0 split)
    CBX1     < 3.557461    to the left,  agree=0.988, adj=0.250, (0 split)
...
```

```
Node number 21: 22 observations,    complexity param=0.02073444
  events=12,  estimated rate=1.053451 , mean deviance=1.249303
  left son=42 (9 obs) right son=43 (13 obs)
  Primary splits:
    RBBP6     < 6.701394   to the right, improve=14.876660, (0 missing)
    LINC00674 < 1.226474   to the right, improve=10.274210, (0 missing)
    RAN       < 5.256385   to the left,  improve= 9.251677, (0 missing)
    PCNT      < 7.818738   to the right, improve= 9.048840, (0 missing)
    EXOG      < 3.538029   to the right, improve= 9.012937, (0 missing)
  Surrogate splits:
    POLI      < 4.773786   to the right, agree=0.909, adj=0.778, (0 split)
    GLTSCR1L  < 6.026291   to the right, agree=0.864, adj=0.667, (0 split)
    EXOG      < 3.581855   to the right, agree=0.864, adj=0.667, (0 split)
    DDX26B    < 5.324905   to the right, agree=0.864, adj=0.667, (0 split)
    CHD9      < 8.454601   to the right, agree=0.864, adj=0.667, (0 split)
```

```
Node number 22: 9 observations
  events=2,  estimated rate=0.3765748 , mean deviance=1.164609
```

```
Node number 23: 141 observations
  events=106,  estimated rate=2.04031 , mean deviance=0.9502795
```

```
Node number 36: 38 observations
  events=9,  estimated rate=0.3453776 , mean deviance=0.907443
```

```
Node number 37: 187 observations
  events=100,  estimated rate=1.030417 , mean deviance=1.237567

Node number 42: 9 observations
  events=1,  estimated rate=0.2522152 , mean deviance=0.6647649

Node number 43: 13 observations
  events=11,  estimated rate=2.217843 , mean deviance=0.5499571

n= 515

node), split, n, deviance, yval
     * denotes terminal node

 1) root 515 692.198200 1.00000000
   2) FCHSD2>=5.794435 507 651.221500 0.97122480
     4) FAM122A>=2.87874 319 394.456500 0.74893350
       8) SRSF9< 4.5731 87  73.027710 0.37756740 *
       9) SRSF9>=4.5731 232 303.039000 0.92130140
        18) CCL22>=-2.369132 225 281.326600 0.87320710
          36) CPOX< 3.77477 38  34.482840 0.34537760 *
          37) CPOX>=3.77477 187 231.425000 1.03041700 *
        19) CCL22< -2.369132 7   7.179567 3.95675300 *
     5) FAM122A< 2.87874 188 228.217600 1.46876800
      10) TLR1>=5.060324 38  44.443570 0.54229060
        20) RUVBL2>=4.298008 16   1.841672 0.07916409 *
        21) RUVBL2< 4.298008 22  27.484670 1.05345100
          42) RBBP6>=6.701394 9   5.982884 0.25221520 *
          43) RBBP6< 6.701394 13   7.149442 2.21784300 *
      11) TLR1< 5.060324 150 159.817400 1.83472200
        22) MED24< 5.218285 9  10.481480 0.37657480 *
        23) MED24>=5.218285 141 133.989400 2.04031000 *
   3) FCHSD2< 5.794435 8  14.485010 6.63701600 *
```

One can also use the R function `rpart.plot(rpart.obj)` to show the recursive partitions graphically (Figure 4.3). As a general practice, a large tree is difficult for users to grasp the essence even though it shows lots of details. One can prune the tree to a proper size using the value of `Cp` for that purpose. There are other methods to control the tree size in the R package manual.

4.2   Random Forest

Like any statistical analysis, one would like to optimize the accuracy and precision of the estimate, namely, to minimize the bias and variance. In the

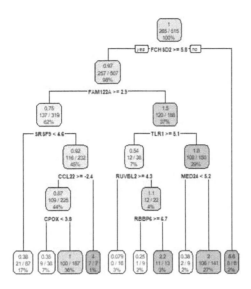

FIGURE 4.3
Graphical exhibition of overall survival recursive partitioning.

data analysis using tree-based models, one can also try to achieve this goal by using Random Forest instead of just using the estimate from a single tree.

As the name indicates, this technique creates many trees, sometimes in hundreds, to make the collection of trees into a forest, and conduct the prediction from regression or classification based on the aggregated results from these trees.

For example, in the classification settings, the random forests grow many classification trees. To classify a new case, one inputs the data of the new case through each of the trees in the forest. Each tree gives a classification, the forest chooses the class for the new case using the class with the highest frequency among all classes it gets classified from all the trees it ran through.

The trees in random forest are grown based on the following criteria:

1. if the number of cases in the training set is N, a bootstrap sample of N cases is created from the original data. This sample will be the training set for growing the tree;

2. if there are M input variables, one samples m variables at random from these M variables with $m << M$ and held the value m fixed throughout the tree growing step. These m variables are used to create the best split at each node; and

3. each tree is grown to the largest extent possible without pruning.

It had been shown that the forest error rate depends on the correlation between any two trees in the forest. Higher correlation usually increases the forest error rate. And the error rate of each individual tree in the forest also affects the forest error rate, higher strength of the individual trees lowers the forest error rate. Reducing m reduces both the correlation and the strength and increasing m increases both. Using the out-of-bag (OOB), which will be explained later, error rate, an "optimal" range of m can be found.

4.2.1 Mechanism of Action of Random Forests

When a bootstrap sample is drawn from the original sample, generally, approximately one-third of the cases are left out of the original sample. This left-out data set, the out-of-bag (OOB) data set, is used to obtain an unbiased estimate of the classification error as trees are added to the forest.

Specifically, after the construction of the kth tree, each left-out case is put through the kth tree to get a classification. In this way, a test set classification is obtained for each case in about one-third of the trees. At the end of the runs, the case is classified as the class which gets the most of the "votes." And the proportion of times the case did not equal to the true class is estimate of the OOB error rate.

This has proven to be unbiased in many tests. Therefore, in random forests, there is no need for cross-validation or a separate test set to get an unbiased estimate of the test set error. Bagged classifiers in Breiman (1996b) give empirical evidence to show that the out-of-bag estimate is as accurate as using a test set of the same size as the training set. Therefore, using the out-of-bag error estimate removes the need for a set aside test set.

Data sets with many weak inputs are becoming more common, e.g., in medical diagnosis, document retrieval, etc. The common characteristics are no single input or small group of inputs can distinguish between the classes. This type of data is difficult for the usual classifiers such as neural nets or trees. Based on the empirical studies, random forests seem to have the ability to work with very weak classifiers as long as their correlation is low.

4.2.2 Variable Importance

In addition to the estimate of the unbiased error rate, random forests can also be used to get estimates of importance for variables in the sample.

Specifically, for every tree in the forest, run the OOB cases through the trees and count the number of correct classification. Consider a variable, e.g., variable m, in OOB cases and randomly permute the values of variable m among the OOB cases, then run these OOB cases with permuted variable m though the forest and count the number of correct classification. The difference of number of correct classification between the pre-permute and post-permute

OOB cases is an indication of how much the classification is affected by variable m, and hence the importance of variable m.

The average of this number over all trees in the forest is the score of importance for variable m. If the values of this score from tree to tree are independent, then the standard error and the z-score can be computed by the conventional computation. The importance of variables considered can be plotted graphically with their score of importance, which can be either mean-squared error or Gini index.

4.2.3   Random Forests for Regression

Random forests for regression are formed by growing trees using the data independently sampled from certain distribution such that the tree predictor takes on numerical values instead of the categorical class labels. The output values are numerical and the mean-squared errors are used to access the goodness of fit of the predictor to the data responses. The random forest predictor is formed by taking the average of the predictions over all the trees created. Random inputs and random features produce good results in classification; however, it is less so in regression.

4.2.4   Example of Random Forest Data Analysis

The same data analyzed using recursive partitioning **rpart** in the previous section are also analyzed using **randomForest** to investigate the efficiency and error rate of using the randomForest method. The R-codes and results are shown below and graphical display of variable importance are also shown next.

4.2.4.1   randomForest for Binary Data

The same TCGA data with binary outcome analyzed by **rpart** in the previous sections are also used for **randomForest** data analysis:

```
library(randomForest}

y<-factor(TCGA4Tree$GradeN)   # the tumor grade data
x<-as.matrix(TCGA4Tree[,-c(1:6)])   # the gene sequence data
rfTCGA<-randomForest(x,y, importance=TRUE)
print(rfTCGA)

Call: randomForest(x = x, y = y, importance = TRUE)
               Type of random forest: classification
                     Number of trees: 500
No. of variables tried at each split: 135

# Shows the out-of-bag error rate and the Confusion-Matrix of the
```

goodness of classification of each category.

OOB estimate of error rate: 32.2%

Confusion matrix:
   0   1 class.error
0 2 50   0.9615385
1 7 118   0.0560000

To show the important variables based on either MSE or Gini Index
  criteria.
  MSE is for regression tree and Gini Index is for classification.
  Print the 20 most important variables only:

MSE<-t(t(tail(importance(rfTCGA)[order(importance(rfTCGA)[,3]),3],20)))
colnames(MSE)<-c("MSE");
print(MSE)
                 MSE
CNNM1   1.550322
IRF3   1.562562
MAP1LC3A   1.572528
POLR2K   1.573244
PEF1   1.576827
KIR2DL4   1.595154
NUP54   1.621320
LHX2   1.655753
ACCS   1.667802
TRIM34   1.668240
SLC7A13   1.672440
TRIM56   1.695259
C6ORF223   1.714213
FAM155B   1.734127
LINC01565 1.735275
JADE1   1.735498
MROH5   1.736099
TMEM40   1.829389
CHEK1   1.895947
PCDHA8   2.204930

Gini<-t(t(tail(importance(rfTCGA)[order(importance(rfTCGA)[,4]),4],20)))
colnames(Gini)<-c("Gini Index");
print(Gini)
          Gini Index
ALDH1L1   0.06985688
ADAM32   0.07056311
GJB6   0.07077290
CHEK1   0.07120230
TRIM56   0.07373026

```
RAX        0.07408018
FGF13      0.07457849
MCM2       0.07687784
SLC30A10   0.07692239
NUP54      0.08051383
HIST1H3I   0.08137552
PPP2R5B    0.08579525
TMEM40     0.09585773
KLHDC7A    0.09684710
PCDHA8     0.10725822
CTSV       0.10738983
ARHGAP23   0.10748868
TRIM67     0.11095786
FAM155B    0.12080214
EXOC4      0.14590667
```

One can also use the R function `varImpPlot(rfTCGA)` to show the error rate during the bootstrap forest building and the important variables selected by randomForest (Figures 4.4 and 4.5). Since the important variables based on MSE or Gini index may not be identical, users need to be aware the type of analysis, whether it is for classification or regression.

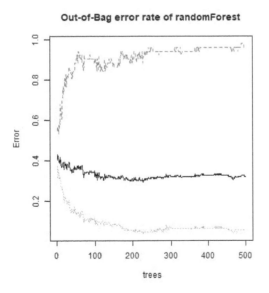

FIGURE 4.4
OOB error rate from randomForest.

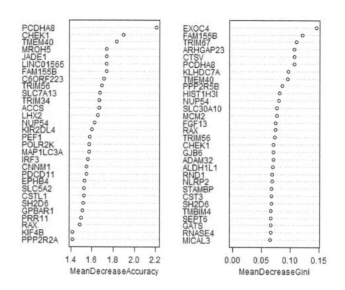

FIGURE 4.5
Important variables (vimp) selected by randomForest based on MSE or Gini index.

4.2.4.2 randomForest for Continuous Data

The Riboflavin production with Bacillus subtilis data, as described in Chapter 1, was also used for the randomForest data analysis.

```
library(randomForest}

rfRIBO<-randomForest(x,y, importance=TRUE)
print(rfRIBO)
plot(rfRIBO, main="OOB prediction error for riboflavin production")
varImpPlot(rfRIBO)

Call: randomForest(x = x, y = y, importance = TRUE)
               Type of random forest: regression
                     Number of trees: 500
No. of variables tried at each split: 1362

          Mean of squared residuals: 0.09302062
                    % Var explained: 62.43
```

One can also use the R function `varImpPlot(rfRIBO)` to show the error rate for the number of trees and the important variables selected by randomForest (Figures 4.6 and 4.7). Even though the important variables based on MSE or Gini index are both shown here, since this is a regression analysis, the MSE panel of Figure 4.7 should be referenced to select the important variables.

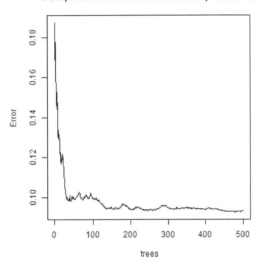

FIGURE 4.6
OOB error rate from randomForest.

4.3 Random Survival Forest

Ishwaran et al. [40] introduced random survival forest (RSF) as an ensemble tree method for analysis of right-censored survival data. Random survival forest methodology extends the Breiman's random forest (RF) method to analyze survival data. RSF strictly follows the principles described by Breiman [10]. In right-censored survival settings, this comprises survival time and censoring status. Thus, the splitting criterion used in growing a tree must explicitly involve survival time and censoring information. Tree node impurity, measuring effectiveness of a split in separating data, must measure separation by survival difference. Further, the predicted value for a terminal node in a tree,

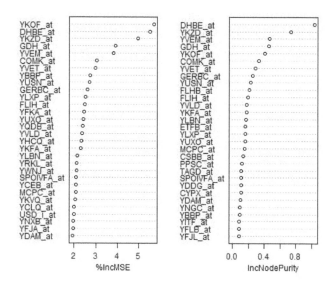

FIGURE 4.7
Important variables (vimp) selected by randomForest based on MSE or Gini index.

the resulting ensemble predicted value from the forest, and the measure of prediction accuracy must all properly incorporate survival information.

4.3.1   Algorithm to Construct RSF

It is important to note that RSF is based on binary tree principles, similar to CART, survival trees are binary trees grown by recursive splitting of tree nodes. A tree is grown starting at the root node comprising all the data. Using a predetermined survival criterion, the root node is split into two daughter-nodes, in turn, each daughter node is split in the same way. The process is repeated in a recursive fashion for each subsequent node. A good split for a node maximizes survival difference between daughters. The best split for a node is found by searching over all possible x variables and split values c, and choosing the x and c that maximizes survival difference.

Similar to random forest, the algorithm for random survival forests can be described in the following steps: (1) draw B bootstrap samples from the original data and (2) grow a survival tree for each bootstrap sample. At each node of the tree, randomly select p candidate variables. The node is split using the candidate variable that maximizes survival difference between descending daughter nodes, then (3) grow the tree to full size under the constraint that a terminal node should have $n \geq d_0 > 0$ unique deaths, = (4) calculate a cumulative hazard function for each tree and average them to obtain the ensemble cumulative hazard function, and finally, (5) uses the out-of-bag data from bootstrap data in step (1) to calculate prediction error for the ensemble cumulative hazard function.

4.3.2 Individual and Ensemble Estimate at Terminal Nodes

At each terminal node, the RSF estimates the cumulative hazard function (CHF) for the subjects fall into the same terminal node by the Nelson-Aalen estimate. Specifically, let $\{t_1 < t_2 < \cdots < t_n\}$ be the distinct event times of the subjects in the terminal node. Define d_{t_i} and Y_{t_i} to be the number of deaths and individuals at risk at time t_i. The Nelson–Aalen estimate of CHF for this terminal node is

$$\hat{H}_t = \sum_{t_i \leq t} d_{t_i}/Y_{t_i}.$$

The ensemble CHF estimate is derived from the individual tree. To compute an ensemble CHF, one can average over B survival trees since each tree in the forest is grown using an independent bootstrap sample. Specifically, it can be estimated by

$$H_e(t|x_i) = \sum_{b}^{B}\{I_{i,b} \times H_b(t|x_i)\}\bigg/ \sum_{b}^{B} I_{i,b}$$

4.3.3 VIMP

Besides various capability of RSF functions, one useful feature of RSF procedure is the capability to estimate the variable importance (vimp). To calculate vimp for a variable x, one can permute the variable among the observations and estimate how much prediction errors changed. More important values indicate variables with predictive ability, whereas zero or negative values identify non-predictive variables. This can be useful in selecting the more important variables for further analysis, especially if the number of variables (e.g., genes for RNA sequence data) is very large.

4.3.4 Example

Using a clinical data set and the corresponding RNA sequence of gene expression, the importance of genes to predict overall survival of the study patients is ordered for all genes based on the procedures described in the random survival method. Given the large number of genes, only the first 18 genes with vimp index greater than 1.0e-4 are shown here. The following program shows the R-codes to produce the vimp and graph (Figure 4.8).

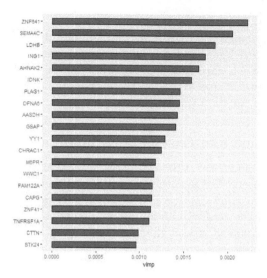

FIGURE 4.8

vimp of RNA for OS - rfsrc.

```
## vimp of genes in prediction of overall survival:

library(randomForestSRC)
library(ggRandomForests)

xx<-RNA2OS  # this data consists of significant genes on OS
xx<-cbind(CNSR, AVAL, RNA2OS)   # combine censoring status, time
and genes.
xx<-data.frame(xx)

surv.f <- as.formula(Surv(AVAL, CNSR) ~ .)
rfsrc.obj <- rfsrc(surv.f, xx, ntree = 150, importance=TRUE)
print(rfsrc.obj)

                          Sample size: 515
                     Number of deaths: 265
                      Number of trees: 150
            Forest terminal node size: 15
        Average no. of terminal nodes: 37.9
No. of variables tried at each split: 21
               Total no. of variables: 406
          Resampling used to grow trees: swor
    Resample size used to grow trees: 325
                             Analysis: RSF
                               Family: surv
                        Splitting rule: logrank *random*
          Number of random split points: 10
```

Error rate: 38.23%

```
vimp2<-t(t(rfsrc.obj$importance)); print(vimp2)
```

Gene	vimp index
AHNAK2	2.748821e-03
IDNK	1.645241e-03
LDHB	1.530274e-03
SEMA4C	1.513145e-03
ZNF641	1.455730e-03
GSAP	1.435533e-03
CXXC5	1.278156e-03
YY1	1.259998e-03
PLAG1	1.249906e-03
ING1	1.213559e-03
CTTN	1.129028e-03
PROSER2	1.112848e-03
PACSIN3	1.078508e-03
NFRKB	1.077702e-03
DFNA5	1.063985e-03
CAPG	1.028287e-03
TTC13	1.012306e-03
WWC1	1.004937e-03
SFPQ	9.672973e-04
CCBL1	9.546484e-04

One can also plot the important variables using `print(plot(gg_vimp(rfsrc.obj, nvar=20)))`, (e.g., only the first 20 variables in this commend).

4.4   XGBoost: A Tree Boosting System

Let $D = \{(y_i, x_i) \mid i = 1, 2, \cdots, n; X_i \in R^m; y_i \in R\}$ be a sample of observed data, and let q represent the structure of a tree that maps each observation of the sample to a leaf of the tree with T leaves.

At each node of tree q, let w denote the score of the node (e.g., in classification tree, w could be the estimated probability of a category; in regression tree, w could be the estimated summary statistics).

With an ensemble of K trees, the outcome of data point x_i can be estimated or predicted by

$$\hat{y}_i = \phi(x_i) = \sum_{k=1}^{K} f_k(x_i), \ f_k \in \mathcal{T}$$

where $\mathcal{T} = \{f(x) = w(q(x))\}$ is the space of regression or classification trees.

To construct a parsimonious tree and not over-fitting, Chen and Guestrin (2016) proposed to minimize the following objective function:

$$\mathcal{L}(\phi) = \sum_i l(y_i, \hat{y}_i) + \sum_k \Omega(f_k), \qquad (4.8)$$

where $l(\cdot)$ is the loss function such as residual sum of squares and $\Omega(f) = \gamma T + (1/2)\lambda \|w\|^2$ representing the penalize term.

To estimate the parameters of equation (4.8), the authors considered the following procedures. At the t-th iteration, let $\hat{y}_i(t)$ be the prediction of the ith observation, they minimized the following objective function:

$$\mathcal{L}^{(t)}(\phi) = \sum_i l(y_i, \hat{y}_i(t-1) + f_t(x_i)) + \Omega(f_t). \qquad (4.9)$$

Equation (4.9) can be approximated with a second-order expansion

$$\mathcal{L}^{(t)}(\phi) \approx \sum_i (l(y_i, \hat{y}_i(t-1)) + g_i f_t(x_i) + (1/2)h_i f_t^2(x_i)) + \Omega(f_t). \qquad (4.10)$$

where $g_i = \partial_{\hat{y}(t-1)} l(y_i, \hat{y}(t-1))$ and $h_i = \partial^2_{\hat{y}(t-1)} l(y_i, \hat{y}(t-1))$.

This can be further simplified as

$$\mathcal{L}^{(t)}(\phi) = \sum_i \{g_i f_t(x_i) + (1/2)h_i f_t^2(x_i)\} + \Omega(f_t). \qquad (4.11)$$

Let $I_j = \{i \mid q(x_i) = j\}$ be the observations in leaf j, then equation (4.11) can be divided by leaves as follows

$$\mathcal{L}^{(t)}(\phi) = \sum_{j=1}^{T} \{\sum_{i \in I_j} g_i w_j + (1/2)\sum_{i \in I_j}(h_i + \lambda)w_j^2\} + \gamma T. \qquad (4.12)$$

To optimize the tree structure by minimizing the impurity of the tree, one can minimizing the objective function in equation (4.12) by taking derivative with respect to w_j and obtain the optimize w_j as

$$w_j^* = \sum_{i \in I_j} g_i / \sum_{i \in I_j}(h_i + \lambda)$$

and obtain the minimized objective function as

$$\mathcal{L}^{(t*)}(\phi) = -(1/2)\sum_{j=1}^{T} \{(\sum_{i \in I_j} g_i)^2 / \sum_{i \in I_j}(h_i + \lambda)\} + \gamma T.$$

4.4.1   Example Using xgboost for Data Analysis

The following shows the R-codes to analyze the Indian prostate data using the recursive partitioning method and draw the tree. One can easily see the binary partition values for the variables used. It is also important not to create a tree with too many levels. Large tree usually increases the difficulty for interpretation and over-fitting the data.

The pancreatic cancer gene sequence data set in TCGA and the baseline tumor grade of patients are also analyzed here using the xgboost method. The following shows the R-codes to analyze the data and the outputs, including the graphical exhibition of the important variable.

4.4.1.1   xgboost for Binary Data

The same TCGA data used for **rpart** binary data analysis are also used here for this data analysis.

```
library(xgboost)
y<-as.numeric(TCGA4Tree$GradeN) #make y as numeric
x<-as.matrix(TCGA4Tree[,-c(1:6)])

dtrain<-xgb.DMatrix(x,label=y)
param <- list(max_depth = 4, eta = 1, verbose = 0, nthread = 2,
          objective = "binary:logistic", eval_metric = "error")
# if no new data availablke, use original data to evaluate the goodness
of fit.
watchlist <- list(train = dtrain, eval = dtrain)
bst <- xgb.train(param, dtrain, nrounds = 2, watchlist)
print(bst)

call: xgb.train(params = param, data = dtrain, nrounds = 2,
      watchlist = watchlist)
params (as set within xgb.train):
  max_depth = "4", eta = "1", verbose = "0", nthread = "2",
  objective = "binary:logistic",
  eval_metric = "error", validate_parameters = "TRUE"

number of features: 18271
niter: 2

# evaluation_log (the error rate in training and validation with
new data):

 iter train_error eval_error
    1    0.090395   0.090395
    2    0.011299   0.011299

print(xgb.importance(model = bst))

    Feature        Gain       Cover  Frequency
```

```
 1: ARHGAP23 0.157518425 0.168543229 0.06666667
 2:    TFAP2B 0.113426624 0.146642131 0.06666667
 3:    OR52D1 0.092233272 0.131406585 0.06666667
 4:     PWRN2 0.089933934 0.117123261 0.06666667
 5:    SLC4A2 0.084695913 0.021901098 0.06666667
 6:      GRK4 0.080251341 0.102603341 0.06666667
 7:   FAM159A 0.074826589 0.014283324 0.06666667
 8:    FAM71A 0.072326473 0.015235546 0.06666667
 9:    NLRP10 0.059795935 0.094093529 0.06666667
10:   ZFYVE26 0.055902885 0.021388524 0.06666667
11:    N6AMT1 0.036255976 0.072705004 0.06666667
12: ANKRD34B 0.034399736 0.008509814 0.06666667
13:    TRIM67 0.022712266 0.063468843 0.06666667
14:   ANAPC11 0.016276133 0.009236161 0.06666667
15:     CPXM1 0.009444498 0.012859609 0.06666667
```

One can also use the R function xgb.plot.importance(xgb.importance
(model = bst)) to show the important variables selected by xgboost
(Figure 4.9). Comparing the important variables selected by rpart,
randomForest, xgboost, one can notice that they did not select all the same
set of variables. This is due to both the difference of estimation procedures
and sampling. However, their outputs serve as a starting point to be further
investigated by the analysts and subject-matter experts to find a plausible
explanation.

4.4.1.2   xgboost for Continuous Data

The Riboflavin production with Bacillus subtilis data are also used
here for this data analysis.

```
library(xgboost)


#use riboflavin data
data(riboflavin)
Gene<-riboflavin[,1]; Gene<-Gene[-1]
ribof1<-riboflavin[,-1]
ribof2<-ribof1[-1,]
ribof3<-data.frame(t(ribof2))
colnames(ribof3)<-Gene
x <- as.matrix(ribof3[,-1])
y <- as.numeric(ribof3[,1])

dtrain<-xgb.DMatrix(x,label=y)
param <- list(max_depth = 4, eta = 1, verbose = 0, nthread = 2,
        objective = "reg:squarederror", eval_metric = "error")
# if no new data, use original data to evaluate the goodness of fit
watchlist <- list(train = dtrain, eval = dtrain)
```

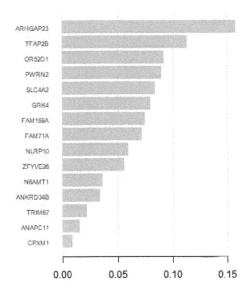

FIGURE 4.9
vimp of genes related to tumor grade.

```
bst <- xgb.train(param, dtrain, nrounds = 2, watchlist)
print(bst)

call: xgb.train(params = param, data = dtrain, nrounds = 2, watchlist
= watchlist) params (as set within xgb.train):
  max_depth = "4", eta = "1", verbose = "0", nthread = "2",
  objective = "reg:squarederror", eval_metric = "error",
  validate_parameters = "TRUE"

niter: 2
nfeatures : 4087
evaluation_log:
  iter train_error eval_error
    1   -5.885159  -5.885159
    2   -5.885159  -5.885159

print(xgb.importance(model = bst))
```

```
   Feature            Gain       Cover  Frequency
1: DHBE_at  0.7033521828  0.27952756      0.125
2: YKOF_at  0.1924329924  0.23622047      0.125
3: ABFA_at  0.0463462161  0.16141732      0.125
4: YUAA_at  0.0333013205  0.14173228      0.125
5: MEND_at  0.0195016314  0.07480315      0.125
6: YXKA_at  0.0039515924  0.04330709      0.125
7:  ABH_at  0.0006820524  0.04330709      0.125
8: AAPA_at  0.0004320119  0.01968504      0.125
```

One can also use the R function `xgb.plot.importance(xgb.importance
(model = bst))` to show the important variables selected by xgboost
(Figure 4.10). Comparing the important variables selected by `rpart`,
`randomForest, xgboost`, one can notice that they did not select all the same
set of variables. This is due to both the difference of estimation procedures
and sampling. However, their outputs serve as a starting point to be further
investigated by the analysts and subject-matter experts to find a plausible
explanation.

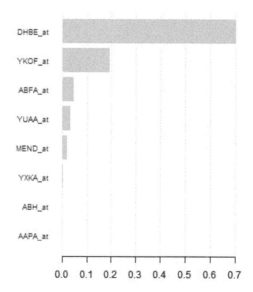

FIGURE 4.10
vimp of genes related to tumor grade.

4.4.2   Example – xgboost for Cox Regression

The same TCGA data used for **rpart** survival data analysis are also used here for this data analysis.

```
library(xgboost)

AVALC<-AVAL * (-1+2*CNSR)  # to put time and censor info together
xx<-cbind(AVALC, RNAseq)
xx<-xx[complete.cases(xx),]
xx<-as.matrix(xx)
x<-xx[,-1]
y<-xx[,1]

dtrain<-xgb.DMatrix(x,label=y)
param <- list(max_depth = 4, eta = 1, verbose = 0, nthread = 2,
         objective = "survival:cox", eval_metric = "cox-nloglik")
# if no new data, use original data to evaluate the goodness of fit
watchlist <- list(train = dtrain, eval = dtrain)  .
bst <- xgb.train(param, dtrain, nrounds = 2, watchlist)
print(xgb.importance(model = bst))
```

	Feature	Gain	Cover	Frequency
1:	PROSER2	0.13536280	0.1273112547	0.0625
2:	CCL22	0.10388819	0.1244315839	0.0625
3:	CTD-3088G3.8	0.10253739	0.0018479805	0.0625
4:	HMGA1	0.10248442	0.1263228293	0.0625
5:	SLC2A1	0.08745274	0.0249826747	0.0625
6:	ING1	0.06657230	0.1203892864	0.0625
7:	DPYSL4	0.06243602	0.1225836060	0.0625
8:	SCN11A	0.05675082	0.0059335374	0.0625
9:	CXXC5	0.05588753	0.1156432826	0.0625
10:	SLC20A2	0.04747568	0.0533576544	0.0625
11:	GCH1	0.04668889	0.0692259516	0.0625
12:	SERTAD3	0.03807668	0.0685323878	0.0625
13:	STC1	0.03502254	0.0283749778	0.0625
14:	PLCD3	0.02373427	0.0047460070	0.0625
15:	UBA6-AS1	0.02285465	0.0053285524	0.0625
16:	CAPG	0.01277508	0.0009884335	0.0625

vimp can also plot using `xgb.plot.importance(xgb.importance(model = bst))` as in Figure 4.11.

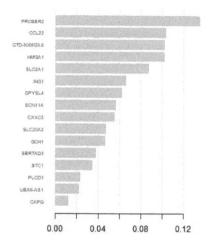

FIGURE 4.11
vimp of genes associated with overall survival.

4.5   Model-based Recursive Partitioning

In many situations of statistical modeling, it is unreasonable to assume that a single global model fits all observations well. But it might be possible to partition the observations with respect to some covariates such that a well-fitting model can be found locally. The recursive partitioning allows for modeling of nonlinear relationships and automated detection of interactions among the explanatory variables.

Several algorithms have been suggested both in the statistical and machine learning communities that attach parametric models to terminal nodes or employ linear combinations to obtain splits in inner nodes. Some of these algorithms (e.g., Loh and Shih 1997) additionally allow one to employ parametric models to obtain splits in inner nodes. Furthermore, maximum likelihood trees (Su et al. 2004) embed regression trees with a constant fit in each terminal node into maximum likelihood estimation.

A segmented parametric model (Zeileis et al. 2008) is proposed and fitted by computing a tree in which every leaf is associated with a fitted model such as maximum likelihood model or a linear regression. In order to optimize the prediction or classification, an objective function of the model is used for estimating the parameters and the split points, and the corresponding model scores are tested for parameter instability in each node to assess which variable should be used for further partitioning.

4.5.1 The Recursive Partitioning Algorithm

The basic idea is that each node is associated with a single model. To assess whether splitting of the node is necessary, a fluctuation test for parameter instability is performed. If there is significant instability with respect to any of the partitioning variables, then split the node into some locally optimal segments and repeat the procedure. If no more significant instabilities can be found, the recursion stops and returns a tree where each terminal node is associated with a model. Two plausible strategies to split a node would be either to use binary split, or to determine the number of daughter-nodes to split adaptively for numerical variables, while always use the number of levels to split for categorical variables. This very general framework for testing parameter stability is called generalized M-fluctuation test and was established by Zeileis and Hornik (2007). The details of splitting and instability test are described in Zeileis et al. [103].

4.5.2 Example

To analyze the relationship between lpsa and lcavol (in the **prostate** data set of **ElemStatLearn** package) conditional on the rest of the variables, one can use the following codes to perform the model-based analysis mob and produce the relationship between the different subgroup of the data. The outputs also provide the conditional model at each node of the tree. For example, the following outputs also show the models for nodes #2 and #3.

```
library(ElemStatLearn)

mobout<-mob(lpsa ~ lcavol | lcp + lweight + age + lbph + svi +
            gleason +  pgg45, data=prostate)
print(summary(mobout))

At node #2: (to specifically examine the output in node #2)

Coefficients:
            Estimate Std. Error t value Pr(>|t|)
(Intercept)  1.53970    0.11734   13.12  < 2e-16 ***
lcavol       0.58640    0.07979    7.35 2.18e-10 ***
---
Signif. codes:  0 '***' 0.001 '**' 0.01 '*' 0.05 '.' 0.1 ' ' 1
(Dispersion parameter for gaussian family taken to be 0.5451635)
 Null deviance: 69.790  on 75  degrees of freedom
Residual deviance: 40.342  on 74  degrees of freedom
```

```
AIC: Inf
```

At node #3: (to specifically examine the output in node #3)

```
Coefficients:
             Estimate Std. Error t value Pr(>|t|)
(Intercept)    2.0536     0.7348   2.795   0.0116 *
lcavol         0.6512     0.2789   2.335   0.0307 *
---
Signif. codes:  0 '***' 0.001 '**' 0.01 '*' 0.05 '.' 0.1 ' ' 1
(Dispersion parameter for gaussian family taken to be 0.700053)
Null deviance: 17.117  on 20  degrees of freedom
Residual deviance: 13.301  on 19  degrees of freedom
AIC: Inf
```

To display the models for different subgroup of data, one can use the following codes, which make it easier to detect the differences between the separate nodes of the relationship between variables or distributions.

```
plot(mobout)
plot(mobout, terminal_panel = node_density)
```

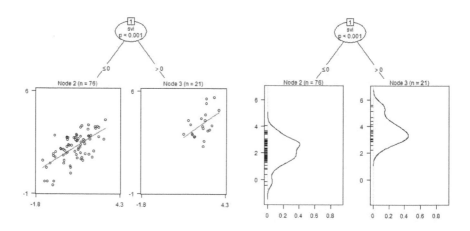

FIGURE 4.12
Model-based modeling.

4.6 Recursive Partition for Longitudinal Data

Longitudinal studies with repeated measures of outcomes of interest had played important roles in many scientific research. The primary objectives are to track the trajectory of the outcomes and to investigate whether any trends of interest existing in overall population or sub-populations and further identify the factors which may cause the differences. Recursive partitioning of data has been a useful tool; therefore, it is quite desirable to have similar tools for longitudinal data.

Longitudinal regression trees were first proposed by Sega (1992) [74], who suggested minimizing within-node Mahalanobis distance of longitudinal response vectors. Yu and Lambert (1999) [101] had further extended that by using smoothing splines to maximize within-node homogeneity of the smoothed trajectories. A R package `splinetree` was created by Neufeld and Heggeseth based on these methodologies.

In `splinetree`, the longitudinal response vector for each individual in the dataset is projected onto a spline basis and created a set of coefficients for each individual, then the coefficients from the projection are used in constructing a regression tree that maximizes the within-node homogeneity of the projected trajectories.

4.6.1 Methodology

Specifically, for each subject i at time $t(1 \leq t \leq q)t$, consider the following model:

$$Y_i(t) = f_i(t) + \epsilon_i(t) = \sum_{k=1}^{q} \beta_{ik} X_k(t) + \epsilon_i(t)$$

for a set of basis functions $X_k(t)$ and coefficient vector $\beta_i = (\beta_{i1}, \cdots, \beta_{iq})$ with $\epsilon_i(t) \sim N(0, \sigma)$.

Therefore, given the observed data, each subject is associated with a set of projection coefficients $\beta_i = (\beta_{i1}, \cdots, \beta_{iq})$. In the regression tree, terminal nodes are labeled with average coefficient vectors $\bar{\beta} = (\bar{\beta}_1, \cdots, \bar{\beta}_q)$ instead of the average of the original data of the subjects in the same node. These average coefficients, along with the basis functions $X_k(t)$, describe the average trajectory for individuals whose covariate values place them in this node.

To split a node, `splinetree` use a procedure similar to CART (and hence `rpart`). At each node the splinetree algorithm performs a greedy search for to the possible values of a variable to split based on the smoothed response curves and to reduce the sum of squared errors (SS) of the smoothed trajectories in the node around the mean smoothed trajectory in the node, namely,

$$SS = \sum_i (X\beta_i - X\bar{\beta}_i)^t (X\beta_i - X\bar{\beta}_i).$$

The errors are evaluated at a set of fixed grid points.

4.6.2   Recursive Partition for Longitudinal Data Based on Baseline Covariates

In scientific research evaluating the effect of interventions, such as the medical treatment effect, the baseline variables are usually important to be included in the statistical models. This is a common practice in either cross-sectional or longitudinal data analysis. To construct regression trees for longitudinal data with baseline covariates, Kundu et al. [50] proposed a longitudinal tree approach to identify the possible subgroups of subjects characterized by baseline covariates so that subjects in each subgroup will have more homogeneous longitudinal data profiles. (Note: how homogeneity is defined could depend on the contexts of the data analysis.)

4.6.2.1   Methodology

Kundu et al. [50] consider the following linear mixed effects model for the longitudinal responses:

$$y_{it} = \beta_0 + \beta_1 t + w_{it}^T \beta + z_{it}^T b_i + \epsilon_{it}$$

where i is the subject index and y, t and $w = (w_1, \cdots, w_q)$ denote the outcome variable, time, and the vector of measurements of scalar covariates, respectively. In addition, z is the random effect component of the model, and $\epsilon \approx N(0, \cdot)$ is the error term which is independent of b.

Let $(X_1, \cdots, X_s)$ be the candidate baseline variables for splitting during the construction of the regression trees with, e.g., $(X_1^{c_1}, \cdots, X_s^{c_s})$ denoting one of the potential cut-off points at $(c_1, \cdots, c_s)$ for each of the baseline variable based on the unique values of the variables. After the cut-off points are determined, one would then consider the following model which incorporates the cut-off information

$$y_{it} = \beta_0^x + \beta_1^x t + w_{it}^T \beta^x + z_{it}^T b_i + \epsilon_{it}.$$

The unknown parameters can be estimated using the maximum likelihood method. In constructing a longitudinal tree through binary partitioning, one way to choose a partition is via maximizing improvement in a goodness-of-fit criterion.

Instead of using greedy search, the authors propose the LongCART algorithm for construction of a regression tree under the conditional inference framework of regression tree construction suggested by Hothorn et al. [36].

In this framework, the authors first identify whether any partitioning variable is associated with the heterogeneity of response trajectory through formal statistical testing via a global "test for parameter instability," which is to detect any evidence of heterogeneity of model parameters across all of its cut-off points in a partitioning variable.

Parameter instability test is carried out for each partitioning variable separately with an adjustment for testing multiplicity. If one or more partitioning variables are found to be significantly associated with the heterogeneity of the response trajectory, the partitioning variable with the minimum p-value is selected as a splitting variable. Once the splitting variable is chosen, the cut-off point with the maximum improvement in goodness-of-fit criterion is used for binary splitting.

4.6.3 LongCART Algorithm

When more than one partitioning variable is found to be significant at level α based on the parameter instability test, the LongCART selects the partitioning variable with the smallest p-value to split the node. Similar p-value methods have been used in other tree algorithms. The advantage of p-value approach is that it offers unbiased partitioning variable selection when the partitioning variables are measured at different scales. The authors propose the following algorithm to construct a regression tree for longitudinal data.

Step 1. Obtain the instability test's p-value for each partitioning variable separately. If there are multiple partitioning variables, adjust the α level that the p-values are compared to. Step 2. Stop if no partitioning variable is significant at level α. Otherwise, choose the partitioning variable with the smallest p-value and proceed to Step 3. Step 3. Consider all cut-off points of the chosen partitioning variable. At each cut-off point, calculate the improvement in the goodness-of-fit criterion (e.g., AIC) due to splitting. Step 4. Choose the cut-off value that provides the maximum improvement in goodness-of-fit criterion and use this cut-off for binary splitting. Step 5. Follow Steps 1 to 4 for each non-terminal node.

4.6.4 Example of Recursive Partitioning of Longitudinal Data

To illustrate the use of longitudinal partitioning, the ACTG175 clinical trial data are used. ACTG 175 was a randomized clinical trial to compare monotherapy with zidovudine or didanosine with combination therapy with zidovudine and didanosine or zidovudine and zalcitabine in adults infected with the human immunodeficiency virus type I, whose CD4 T cell counts were between 200 and 500 per cubic millimeter. A data frame with 6417 observations from 2139 patients on 24 variables. Details can be found in the literature.

```
library(LongCART)

# Get the data from large scale clinical trial
data(ACTG175)
# specify variables
gvars=c("age", "gender", "wtkg", "hemo", "homo", "drugs", "karnof",
  "oprior", "z30", "zprior", "race", "str2", "symptom", "treat",
  "offtrt")
tgvars=c(1, 0, 1, 0, 0, 0, 1, 0, 0, 0, 0, 0, 0, 0, 0)
out<- LongCART(data=ACTG175, patid="pidnum", fixed=cd4 ~ time, gvars=gvars,
```

```
    tgvars=tgvars, alpha=0.05, minsplit=100, minbucket=50, coef.digits=2)
print(out)

# a detailed variable splitting history for each variable
Splitting variable: age
G=59
Stability Test for Continuous grouping variable
Test.statistic= 1.549 0.877,    Adj. p-value=0.0330.425

Splitting variable: gender
G=2
Stability Test for Categorical grouping variable
Test.statistic=0.658,    p-value=0.72

Splitting variable: wtkg
G=667
Stability Test for Continuous grouping variable
Test.statistic= 0.962 1.455,    Adj. p-value=0.3130.058
...

# to show the statistics of the tree
```

	ID	n	yval	var	index	p (Instability)	loglik	improve	Terminal
1	1	2139	362.79-0.35time	offtrt	1.0	0.000	-35134	171	FALSE
2	2	1363	378.48-0.23time	treat	1.0	0.000	-23699	78	FALSE
3	4	316	372.95-0.75time	str2	1.0	0.039	-5447	37	FALSE
4	8	125	398.96-0.44time	gender	NA	1.000	-2158	NA	TRUE
5	9	191	355.93-0.95time	oprior	NA	1.000	-3269	NA	TRUE
6	5	1047	380.12-0.07time	str2	1.0	0.008	-18211	54	FALSE
7	10	426	404.62+0.04time	wtkg	74.5	0.030	-7465	15	FALSE
8	20	196	401.18-0.23time	age	NA	0.942	-3439	NA	TRUE
9	21	230	407.49+0.27time	wtkg	NA	1.000	-4016	NA	TRUE
10	11	621	363.34-0.15time	age	NA	0.079	-10717	NA	TRUE
11	3	776	338.27-0.9time	z30	1.0	0.002	-11349	46	FALSE
12	6	364	360.77-0.72time	oprior	NA	0.082	-5368	NA	TRUE
13	7	412	318.52-1.08time	age	NA	0.671	-5955	NA	TRUE

```
AIC(tree)=-69900    AIC(Root)=-70276
logLikelihood (tree)=-34922    logLikelihood (Root)=-35134
Deviance=424 (df=12, p-val=0)
```

One can also plot the tree by using `par(xpd = T); plot(out, compress = T, uniform=TRUE); text(out, use.n = T)`. The longitudinal tree is shown in Figure 4.13.

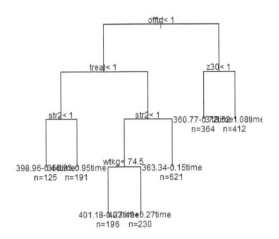

FIGURE 4.13
Recursive partition of longitudinal data.

4.7   Analysis of Ordinal Data

For ordinal response classification, the Gini index as discussed previously can be used to perform the node splitting. In addition to using Gini index, Breiman et al. [9] also proposed another method, named "Twoing," to split the nodes.

Specifically, denote the set of classes by $C = \{1, 2, \cdots, J\}$. At each node t, separate the classes into two superclass,

$$C_1 = \{j_1, \cdots, j_k\}, \ C_2 = C - C_1. \tag{4.13}$$

For a given split s of node t, compute the reduction of impurity, $\Delta(s, t, C_1)$, as though it corresponded to a two-class problem. Find the split s^* which maximizes $\Delta(s, t, C_1)$ among all splits s. By changing the members in C_1 and C_2, one can find a particular superclass C_1^* such that

$$\Delta i(s^*, t, C_1^*) = \max_{s, C_1} (s, t, C_1) \tag{4.14}$$

and the classification can be properly assigned.

Another proposed ordinal impurity function for deriving an ordinal response classification tree based on a measure of nominal-ordinal association (Piccarreta [65]) that does not require the assignment of costs of misclassification is

$$i_p(t) = \sum_{j=1}^{J} F(\delta_j|t)(1 - F(\delta_j|t)) \tag{4.15}$$

where $F(\delta_j|t) = \sum_{k=1}^{j} p(\delta_k|t)$ (Piccarreta [66]).

For nominal response prediction, misclassification rates are often examined as a means for assessing the performance of the classifier. For ordinal response prediction problems, the gamma statistic (Agresti [2]) can be used as an ordinal measure of association between the observed and predicted responses as a means for gauging the success of ordinal classification.

4.8   Examples – Analysis of Ordinal Data

In the following, we analyze the Cleveland Clinc Heart Data and examine the partitioning of response based on the covariates, using the `ordinal` or `twoing` method. Even though the classifications are not identical, they essentially provide similar conclusions.

4.8.1   Analysis of Cleveland Clinic Heart Data (Ordinal)

```
Call:
rpart(formula = HeartDisease ~ Age + Sex + ChestPain + BPS +
    Cholestoral + BloodSugar + EKG + HeartRate + Angina + STdep +
    STslope + ColorVessel + Thal, data = Clevland, method = ordinal)

n= 303
           CP nsplit rel error
1  0.25058300      0 1.0000000
2  0.10633705      1 0.7494170
3  0.05465066      2 0.6430800
4  0.02946367      4 0.5337786
5  0.02499179      5 0.5043150
6  0.02430581      6 0.4793232
7  0.01959998      7 0.4550174
8  0.01841856      8 0.4354174
9  0.01701246      9 0.4169988
10 0.01178423     10 0.3999864
11 0.01000000     11 0.3882021
```

```
node), split, n, deviance, yval
      * denotes terminal node

  1) root 303 455.808600 0
    2) Thal< 4.5 167 115.473100 0
      4) ColorVessel< 0.5 118   20.550850 0 *
      5) ColorVessel>=0.5 49   71.918370 0
       10) ChestPain< 3.5 29   14.551720 0 *
       11) ChestPain>=3.5 20   30.550000 1
         22) Angina< 0.5 9    8.888889 1 *
         23) Angina>=0.5 11   12.727270 3 *
    3) Thal>=4.5 136 226.117600 0
      6) STdep< 0.7 46   41.826090 0
       12) ColorVessel< 0.5 26   11.884620 0 *
       13) ColorVessel>=0.5 20   18.550000 1 *
      7) STdep>=0.7 90 135.822200 3
       14) ChestPain< 1.5 7   13.428570 0 *
       15) ChestPain>=1.5 83 108.963900 3
         30) STdep< 2.35 54   73.333330 3
           60) ColorVessel< 1.5 37   44.702700 1
            120) HeartRate< 148.5 29   29.448280 1
              240) STdep< 1.45 16   13.000000 3 *
              241) STdep>=1.45 13   11.076920 1 *
            121) HeartRate>=148.5 8    7.500000 0 *
           61) ColorVessel>=1.5 17   20.235290 3 *
         31) STdep>=2.35 29   24.551720 2 *
```

(Ref: Figure 4.14)

4.8.2 Analysis of Cleveland Clinic Heart Data (Twoing)

```
Call:
rpart(formula = HeartDisease ~ Age + Sex + ChestPain + BPS +
    Cholestoral + BloodSugar + EKG + HeartRate + Angina + STdep +
    STslope + ColorVessel + Thal, data = Clevland, method = twoing)

n= 303
          CP nsplit rel error
1  0.25058300      0 1.0000000
2  0.09678986      1 0.7494170
3  0.05465066      2 0.6526271
4  0.03025422      4 0.5433258
5  0.02839408      5 0.5130716
6  0.01959998      6 0.4846775
7  0.01548678      7 0.4650775
8  0.01289528      9 0.4341040
9  0.01268175     10 0.4212087
```

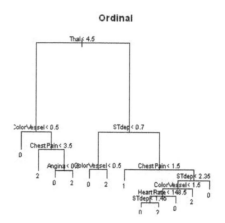

Ordinal

FIGURE 4.14
Analysis of Cleveland Clinic Heart Data.

```
10 0.01000000      11 0.4085269

node), split, n, deviance, yval
      * denotes terminal node

 1) root 303 455.808600 0
   2) Thal< 4.5 167 115.473100 0
     4) ColorVessel< 0.5 118   20.550850 0 *
     5) ColorVessel>=0.5 49   71.918370 0
      10) ChestPain< 3.5 29   14.551720 0 *
      11) ChestPain>=3.5 20   30.550000 1
        22) Angina< 0.5 9    8.888889 0 *
        23) Angina>=0.5 11   12.727270 3 *
   3) Thal>=4.5 136 226.117600 0
     6) STdep< 0.85 51   57.294120 0
      12) Cholestoral< 240.5 27   12.518520 0 *
      13) Cholestoral>=240.5 24   31.833330 1
        26) ChestPain< 3.5 9    5.555556 0 *
        27) ChestPain>=3.5 15   20.400000 1 *
     7) STdep>=0.85 85 124.705900 3
      14) ChestPain< 1.5 7   13.428570 0 *
      15) ChestPain>=1.5 78   97.487180 3
```

```
30) HeartRate< 133.5 38   45.394740 3
  60) BPS< 144.5 28   26.714290 3 *
  61) BPS>=144.5 10   12.900000 4 *
31) HeartRate>=133.5 40   46.775000 2
  62) ColorVessel< 1.5 23   23.739130 2 *
  63) ColorVessel>=1.5 17   14.235290 2 *
```

(Ref: Figure 4.15)

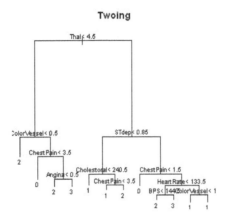

FIGURE 4.15
Analysis of Cleveland Clinic Heart Data.

4.9   Advantages and Disadvantages of Trees

Even though tree-based methods have been widely used nowadays, it is just one of the tools in the data analysis toolbox. It is useful to help analysts get some kind of preliminary understanding of the data structures and potential subgroups. However, it is not a panacea, it also has its advantages and disadvantages. For example, (1) trees are very easy to explain to people. In fact, they are even easier to explain than linear regression, (2) decision trees are more similar to human decision-making than do the regression and classification approaches, (3) it can be displayed graphically, and are easily interpreted even by a non-expert (but beware of the large-sized trees), (4) tree can handle qualitative predictors without the need to create dummy variables such as in regression setting, (5) since this is just a good explanatory tool, model-based analysis conducted at the terminal nodes can help further understanding of the relationships between dependent and independent variables and that can be very helpful.

5

Support Vector Machine

5.1 General Theory of Classification and Regression in Hyperplane

Consider a set of observations

$$(\mathbf{Y}(t), \mathbf{X}(t)) = \{(y_1(t), x_1(t)), (y_2(t), x_2(t)), \cdots, (y_n(t), x_n(t))\} \qquad (5.1)$$

where

$$y_i(t) = \begin{cases} 1 & \text{if } x_i(t) \in \text{class I} \\ 0 & \text{if } x_i(t) \in \text{class II} \end{cases}$$

at time t. A perceptron learning machine was proposed (Rosenblatt, 1962) to learn from observations in neurophysiology.

The learning machine is a binary indicator function f, which is linear function of factors based on the original observations in Equation (5.1), specifically,

$$f(x, \beta(t)) = \text{sign}\left\{ \sum_{p=1}^{n} \beta_p(t) u_p(x) \right\}, \qquad (5.2)$$

where the functions

$$\mathbf{U}(t) = \{u_1(x(t)), u_2(x(t)), \cdots, u_n(x(t))\}$$

maps the x's in sample space to superpositions u's of the feature space, and f is a linear indicator functions in this space which can be constructed using recursive procedures.

To put this in a more general term, let A be a nonlinear operator which maps vectors $x \in \mathbf{X}$ into vector $u \in \mathbf{U}$, a feature space, the perceptron constructs a separating hyperplane

$$f(\mathbf{U}(t), \beta(t)) = \text{sign}\{(\mathbf{U}(t) * \beta(t))\}$$

through the origin in $\mathbf{U}$ space.

The problem of constructing the nonlinear decision rule in the space $\mathbf{X}$ reduces to constructing a separating hyperplane in the space $\mathbf{U}$. In space $\mathbf{U}$, the unknown parameters can be estimated using the following steps iteratively when the number of data increased.

DOI: 10.1201/9781003205685-5

Let $\beta(t) = \{\beta_1(t), \cdots, \beta_n(t)\}$, at each new time point, f can be updated by the following rule with new data

$$\beta(t) = \begin{cases} \beta(t-1), & \text{if } y(t)(\beta(t-1) * u(t)) > 0 \\ \beta(t-1) + y(t)u(t), & \text{if } y(t)(\beta(t-1) * u(t)) \leq 0 \end{cases} \tag{5.3}$$

5.1.1   Separable Case

Now consider a sequence of observations $(Y, \mathbf{U}) = \{(y_1, u_1), \cdots, (y_n, u_n), \cdots\}$, if there is a unit vector β_0 such that

$$\min_{(y,u) \in \{Y, U\}} = y(\beta_0 * U) \geq \rho_0 \tag{5.4}$$

holds true, then Novikoff () proved that under the condition $|u_i|$ is finite, suppose that there exists a hyperplane with coefficients β_0 that separates correctly elements of this sequence and satisfies the condition (5.4), then using the iterative procedure as in equation (5.3), the hyperplane will correctly separates all data points with a finite number of corrections.

5.1.2   Non-separable Case

It is not always possible to construct a hyperplane to separate the training observations without error. That is, there is no w_0 that satisfy the inequality (5.4) for a given small ρ_0. Therefore, one tries to find the vector ρ_0 that provides the local minimum to the risk functional

$$R(w) = \int (y - \text{sign}\{(u * w)\})^2 dP(u, y) \tag{5.5}$$

or the local minimum to the empirical risk functional

$$R_{emp}(w) = \frac{1}{l} \sum_{j=1}^{l} (y_i - \text{sign}\{(u_j * w)\})^2.$$

In the following, we describe two approaches to construct a hyperplane which minimize the risk functional (5.6).

It is not always possible to construct a hyperplane to separate the observations without error. That is, there is no β_0 that satisfy the inequality (5.4) for a given small ρ_0. However, one can try to find a vector ρ_0 that minimizes the risk functional of expected misclassification error, namely,

$$R(\beta) = \int (y - \text{sign}\{(u * \beta)\})^2 dP(u, y) \tag{5.6}$$

over the distribution space of $(\mathbf{Y}, \mathbf{U})$. Empirically, equation (5.6) can be written as

$$R_{emp}(\beta) = \frac{1}{l} \sum_{j=1}^{l} (y_i - \text{sign}\{(u_j * \beta)\})^2. \tag{5.7}$$

Several methods were proposed to minimize the risk functional in (5.6). In the following, we discuss a few commonly used approaches to perform the approximations.

5.1.2.1 Method of Stochastic Approximation

Consider a more general form of (5.6), i.e., let

$$R(\beta) = \int Q(z, \beta) dP(z). \qquad (5.8)$$

In order to minimize the risk functional using the i.i.d. data $\{z_1, \cdots, z_n, \cdots\}$, one can consider the stochastic approximation method using the following recurrent procedure:

$$\beta(t) = \beta(t-1) - \gamma_t \{\text{grad}_\beta Q(z_i, \beta(t-1)) + \xi_i\} \qquad (5.9)$$

where sequence of values $\gamma_t \geq 0$, $\lim_{t\to\infty} \gamma = 0$, $\sum_{t=1}^{\infty} \gamma_t = \infty$, $\sum_{t=1}^{\infty} \gamma_t^2 < \infty$, and $E(\xi|\beta) = 0$. Litvakov (1966) proved that, under certain regularity conditions, that for any initial point β, $R(\beta_i)$ converges to $\inf R(\beta)$ if $i \to \infty$ with probability 1.

5.1.2.2 Method of Sigmoid Approximations

When $Q(z, \beta)$ in equation (5.8) is not differentiable at point z, the stochastic approximation procedures are not applicable. A common alternative is to approximate $Q(z, \beta)$ by a sigmoid function

$$f(u, \beta) = S\{(\beta * u)\}$$

where $S(u)$ is a smooth monotonic function such that $S(-\infty) = -1$ and $S(\infty) = 1$. For example, if $Q(z, \beta) = (z - \text{sign}\{(u * \beta)\})^2$, the gradient can then be estimated by

$$\text{grad}_\beta Q(z, \beta) = \frac{d}{d\beta}[y - S(\beta * u)]^2 = -2[y - S(\beta * u)] \times S'_\beta\{(\beta * u)\}u.$$

With the sigmoid approximation, one can approximate the risk functional (5.7) by

$$R_{emp}(\beta) = (1/n) \sum_{j=1}^{n} (y_j - S(\beta * u_j))^2.$$

The gradient of the empirical risk functional can be estimated as

$$\text{grad}_\beta R_{emp}(\beta) = (-2/n) \sum_{j=1}^{n} (y_j - S(\beta * u_j)) \times S'(\beta * u_j) u_j,$$

and β_t can be updated by

$$\beta_t = \beta_{t-1} - \gamma_t \times \text{grad}_\beta R_{emp}(\beta_{t-1})$$

with $\gamma_t > 0$.

5.1.2.3   Method of Radial Basis Functions

In addition to the sigmoid function approximation, Aizerman et al. (1964) suggested another method of approximation, they proposed to estimate the functional from the data using the following set of functions:

$$f(x, \beta) = \text{sign}\Big\{ \sum_{i=1}^{l} \beta_i \phi(|x - x_i|) \Big\} \beta, \tag{5.10}$$

where $\phi(0) = 1$ and $\lim_{u \to \infty} \phi(|u|) = 0$. Function $\phi(|u|)$, called Radial Basis Functions (RBF), is a monotonic function that converges to zero with increasing $|u|$.

Within the set of RBFs, one minimizes the empirical risk functional

$$R_{emp}(\beta) = \sum_{i=1}^{l} \Big(y_i - \sum_{j=1}^{l} \beta_j \phi(|x_i - x_j|) \Big)^2. \tag{5.11}$$

It was shown that if the matrix $||\phi(|x_i - x_j|)||$ is positive definite, (5.11) has a unique solution for $\min R_{emp}(\beta)$.

5.2   SVM for Indicator Functions

This section discusses the construction of rules to separate two classes of data points in certain "optimal" way so that the data points can be separated as much as possible and to reduce the misclassification error.

5.2.1   Optimal Hyperplane for Separable Data Sets

Definition of Separable Hyperplane   Given data set

$$\begin{cases} (y_1 = 1, x_{11}), \cdots, (y_l = 1, x_{1l}) & \text{if } x_{1i} \in \text{class I,} \\ (y_1 = 0, x_{21}), \cdots, (y_l = 0, x_{2l}) & \text{if } x_{2i} \in \text{class II} \end{cases} \tag{5.12}$$

these two classes of data points are said to be separable by the hyperplane, $(x * \phi) = c$, if there exist a unit vector ϕ ($||\phi|| = 1$) and a constant c such that the inequalities

$$\begin{cases} (x_i * \phi) > c & \text{if } x_i \in \text{class I} \\ (x_i * \phi) < c & \text{if } x_i \in \text{class II} \end{cases} \tag{5.13}$$

hold true.

Definition of Maximal Margin (or Optimal) Hyperplane   Let

$$\begin{cases} c_1(\phi) = \min_{x_i \in I}(x_i * \phi), \\ c_2(\phi) = \max_{x_i \in II}(x_i * \phi), \end{cases} \tag{5.14}$$

$$\rho(\phi) = \{c_1(\phi) - c_2(\phi)\}/2, \text{ and } c_0 = \{c_1(\phi) + c_2(\phi)\}/2.$$

Consider the unit vector ϕ_0 which maximizes the function ρ under the condition of (5.13), namely,

$$\rho(\phi_0) = \max_{\phi, ||\phi||=1} \rho(\phi),$$

then vector (ϕ_0, c) determines the hyperplane that separates x's of class I from class II with the maximal margin $\rho(\phi_0)$. This hyperplane is named as the maximal margin hyperplane or the optimal hyperplane. *Note:* It had been shown that the optimal hyperplane is unique.

5.2.1.1 Constructing the Optimal Hyperplane

One can re-write (5.13) as

$$\begin{cases} (x_i * \psi_0) + b_0 \geq 1 & \text{if } y_i = 1 \\ (x_i * \psi_0) + b_0 \leq 1 & \text{if } y_i = -1 \end{cases} \tag{5.15}$$

with a vector ψ_0 and a constant b_0.

Let

$$\phi_0 = \psi_0 / ||\psi_0||,$$

the margin ρ_0 between the optimal hyperplane and separated vectors is equal to

$$\rho(\phi_0) = \sup_{\phi_0} \frac{1}{2} \left(\min_{i \in I}(x_i * \phi_0) - \max_{i \in II}(x_i * \phi_0) \right) = \frac{1}{|\psi_0|}.$$

The vector ψ_0 with the smallest norm satisfying constraints (5.15) with $b = 0$ defines the optimal hyperplane passing through the origin.

Equation (5.15) can be re-expressed in a simpler form, namely,

$$y_i((x_i * \psi_0) + b) \geq 1, \quad i = 1, \cdots, n. \tag{5.16}$$

Therefore in order to find the optimal hyperplane one has to minimize $|\psi|^2 = (\psi * \psi)$ subject to the linear constraints (5.16).

Using Lagrange function and the multipliers, one can find the saddle point of the Lagrange function

$$L(\psi, b, \alpha) = \frac{1}{2}(\psi * \psi) - \sum_{i=1}^{n} \alpha_i(y_i[(x_i * \psi) + b] - 1), \tag{5.17}$$

where $\alpha_i \geq 0$ are the Lagrange multipliers, by minimizing (5.17) over ψ and b and to maximize it over the Lagrange multipliers $\alpha_i \geq 0$.

After taking the required derivatives with respect to the parameters and solve

$$\begin{cases} \partial L(\psi, b, \alpha)/\partial \psi = 0 \\ \partial L(\psi, b, \alpha)/\partial b = 0 \end{cases}$$

one obtains

$$\begin{cases} \psi = \sum_{i=1}^{l} y_i \alpha_i x_i, \\ \sum_{i=1}^{l} y_i \alpha_i = 0. \end{cases} \tag{5.18}$$

Substitute equation (5.18) into equation (5.17), one has

$$\tilde{L}(\psi, b, \alpha) = \sum_{i=1}^{l} \alpha_i - (1/2) \sum_{i,j=1}^{l} y_i y_j \alpha_i \alpha_j (x_i * x_j). \tag{5.19}$$

To construct the optimal hyperplane one has to find the value of coefficients α_i, denotes it by α_i^0, that maximize the function (5.19) and satisfies

$$\alpha_i^0 \geq 0, \ i = 1, \cdots, l$$

under the constraint $\sum_{i=1}^{l} y_i \alpha_i^0 = 0$. Using these coefficients α_i^0, one obtains the solution

$$\psi_0 = \sum_{i=1}^{l} y_i \alpha_i^0 x_i.$$

One needs to estimate the value of b, denoted by b_0, to maximize the margin and satisfies

$$\alpha_i^0 \Big(y_i[(x_i * \psi_0) + b_0] - 1 \Big) = 0, \quad i = 1, \cdots, l. \tag{5.20}$$

From conditions (5.20) one concludes that nonzero values α_i^0 correspond only to the vectors x_i that satisfy the equality

$$y_i \big((x_i * \psi_0) + b_0 \big) - 1 = 0,$$

therefore the optimal hyperplane has the form

$$f(x, \alpha_0) = \sum_{i=1}^{l} y_i \alpha_i^0 (x_i * x) + b_0 \tag{5.21}$$

with α_i^0 and b_0 as estimated above.

5.2.2 Optimal Hyperplane for Non-Separable Sets

Not all the sample data points can be successfully classified into their respective classes. This section describes the methodologies to construct the hyperplane with minimized misclassification error.

5.2.2.1 Generalization of the Optimal Hyperplane

If the sample data set

$$\{(y_1, x_1), \cdots, (y_l, x_l)\}, \ y \in \{1, -1\}$$

cannot be separated without error by a hyperplane, then there is no pair (ψ, b) such that

$$(\psi * \psi) \leq \frac{1}{\rho^2} = A^2$$

and the inequality

$$y_i((x_i * \psi) + b) \geq 1, \quad i = 1, \cdots, l$$

hold true. In this case, one can attempt to construct a hyperplane that makes the smallest number of misclassification errors.

Let $\{\xi_1, \cdots, \xi_l\}$ be a set of nonnegative variables. One wants to find the hyperplane that provides the minimal number of errors by minimizing the functional

$$\Phi(\xi) = \sum_{i=1}^{l} \xi_i$$

subject to the constraints

$$y_i((x_i * \psi) + b) \geq 1 - \xi_i, \text{ and } \xi_i \geq 0 \text{ for } i = 1, \cdots, l, \tag{5.22}$$

with the constraint

$$(\psi * \psi) \leq A^2. \tag{5.23}$$

Then one needs to minimize the functional

$$\Phi(\xi) = \sum_{i=1}^{l} \xi_i$$

subject to the constraints (5.22) and (5.23). The hyperplane

$$(\psi_0 * x) + b = 0$$

constructed on the basis of the solution of this optimization problem is the optimal hyperplane for non-separable case.

To solve this optimization problem, one needs to find the saddle point of the Lagrangian

$$L(\psi, b, \alpha, \beta, \gamma) = \sum_{i=1}^{l} \xi_i - (1/2)\gamma(A^2 - (\psi * \psi))$$

$$- \sum_{i=1}^{l} \alpha_i(y_i((\psi * x_i) + b) - 1 + \xi_i) - \sum_{i=1}^{l} \beta_i \xi_i, \tag{5.24}$$

namely, to minimize the equation with respect to ψ, b, ξ_i and to maximize with respect to nonnegative multipliers $\alpha_i, \beta_i, \gamma$. Taking the appropriate derivatives and solve

$$\partial L/\partial \psi = 0, \quad \partial L/\partial b = 0, \text{ and } \partial L/\partial \xi_i = 0,$$

one obtains

$$\psi = \frac{1}{\gamma} \sum_{i=1}^{l} \alpha_i y_i x_i, \quad \sum_{i=1}^{l} \alpha_i y_i = 0, \text{ and } \alpha_i + \beta_i = 1.$$

Substituting these equality into Equation (5.24), one obtains the functional

$$\tilde{L}(\alpha, \gamma) = \sum_{i=1}^{l} \alpha_i - \frac{1}{2\gamma} \sum_{i,j=1}^{l} \alpha_i \alpha_j y_i y_j (x_i * x_j) - (\gamma A^2)/2. \qquad (5.25)$$

Maximizing Equation (5.25) with respect to $\alpha_i, \beta_i, \gamma$ under the constraints

$$\sum_{i=1}^{l} \alpha_i y_i = 0, \quad 0 \leq \alpha_i \leq 1, \quad \text{and} \quad \gamma \geq 0,$$

one finds the estimated parameter $\hat{\gamma}$ that maximizes (5.25):

$$\hat{\gamma} = \left(\sum_{i,j=1}^{l} \alpha_i \alpha_j y_i y_j (x_i * x_j) \right)^{1/2} \Big/ A.$$

The vector of parameters $\alpha_0 = (\alpha_1^0, \cdots, \alpha_l^0)$ which maximize (5.25) defines the generalized optimal hyperplane. With the substitution and rearranging the terms, one has

$$f(x) = A \times \left(\sum_{i,j=1}^{l} \alpha_i^0 \alpha_j^0 y_i y_j (x_i * x_j) \right)^{-1/2} \times \sum_{i=1}^{l} \alpha_i^0 y_i (x * x_i) + b. \qquad (5.26)$$

The value of the threshold b is chosen to satisfy the Kuhn-Tucker condition

$$\alpha_i^0 \times \left\{ A \times \left(\sum_{i,j=1}^{l} \alpha_i^0 \alpha_j^0 y_i y_j (x_i * x_j) \right)^{-1/2} \times \sum_{i=1}^{l} \alpha_i^0 y_i (x * x_i) + b \right\} = 0$$

for $i = 1, \cdots, l$.

5.2.3   Support Vector Machine

The separable or non-separable hyperplane as shown in Equations (5.21) and (5.26) is the combinations of the products of $y_i \alpha_i^0$ and the inner products of data from sample space. In support vector machine (SVM), instead of considering the inner products of data in sample space, one considers the inner products in the feature space through some pre-specified nonlinear data mapping via kernel functions which satisfy the Mercer condition. Schematically, the set of support vectors (as a subset of the original data) is transformed nonlinearly and properly weighted by $y_i \alpha_i^0, (i = 1, \cdots, N)$. Values of a function of these weighted quantities are the basis of the decision rules for classification. Graphically, the schematic process can be illustrated as shown in Figure 5.1.

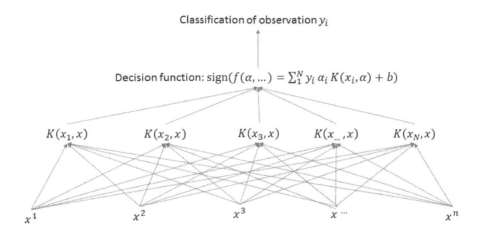

FIGURE 5.1
SVM network.

5.2.4 Constructing SVM

Specifically, suppose one maps the vector $x \in R^n$ into a hyperspace (e.g., Hilbert space) with coordinates $\{z_1(x), z_2(x), \cdots, z_n(x), \cdots\}$, then according to the Hilbert-Schmidt theory the inner product in a Hilbert space has an equivalent representation:

$$(z_1 * z_2) = \sum_{r=1}^{\infty} a_r z_r(x_1) z_r(x_2) \quad a_r \geq 0. \tag{5.27}$$

If a kernel $K(u, v)$ satisfies the Mercer condition, then $K(u, v)$ can be expressed by an asymptotic expansion such as (5.27). With this property, one can construct SVM in the feature space.

In the feature space, instead of maximizing the functional in (5.19) for the separable case, one can maximize

$$\tilde{L}(\alpha) = \sum_{i=1}^{l} \alpha_i - (1/2) \sum_{i,j=1}^{l} y_i y_j \alpha_i \alpha_j K(x_i, x_j), \tag{5.28}$$

and in the non-separable case, instead of maximizing the functional in (5.25), one can maximize

$$\tilde{L}(\alpha) = \sum_{i=1}^{l} \alpha_i - A \times \left\{ \sum_{i,j=1}^{l} \alpha_i \alpha_j y_i y_j K(x_i, x_j) \right\}^{1/2}, \tag{5.29}$$

subject to the same constraints as in those functionals for the sample space.

Generally, any kernels satisfy the Mercer condition can be used in the estimation. In practice, the polynomial kernel and radial basis kernels described below are among the most commonly used.

5.2.4.1   Polynomial Kernel Functions

A polynomial SVM of degree d is defined as

$$K(x, x_i) = [(x * x_i) + 1]^d.$$

Using this kernel, one can construct the kernel function and the decision rule function

$$f(x) = \text{sign}\left(\sum_{i=1}^{N} y_i \alpha_i [(x * x_i) + 1]^d + b \right). \tag{5.30}$$

5.2.4.2   Radial Basis Kernel Functions

Classical radial basis function (RBF) machines use the following set of decision rules:

$$f(x) = \text{sign}\left(\sum_{i=1}^{N} a_i K_\gamma(|x - x_i|) - b \right), \tag{5.31}$$

where $K_\gamma(|x - x_i|)$ depends on the distance $|x - x_i|$ between two vectors. The function $K_\gamma(|z|)$ is a positive definite monotonic function for any fixed γ; it tends to zero as $|z|$ goes to infinity. The most popular function of this type is

$$K_\gamma(|x - x_i|) = \exp\{-\gamma|x - x_i|^2\}.$$

To construct the decision rule from the set (5.31), one has to estimate the number N of the centers x_i, the vectors x_i describing the centers, the values of the parameters a_i, and the value of the parameter γ.

5.2.5   Example: Analysis of Binary Classification Using SVM

In this section, we use the NCI60 data set for the analysis. The cancer types in this data were re-classified into solid tumors and non-solid tumors. The gene data are use as covariates. The purpose is to use SVM to select the important genes to predict the types of cancers.

```
library(e1071)
library(kernlab)

#data(NCI60)
x <- NCI60$data[ , featureRankedList[1:100]]
y <- (cancer2-1)*2-1  ## to make y with 1 or -1 values.

# use svm in linear kernel
svmmodel <- svm(x[ , featureRankedList[1:100]], y, cost = 10, kernel="linear")
print(svmmodel)
# outputs:
```

```
Call: svm.default(x = x[, featureRankedList[1:100]], y = y, kernel = "linear", cost = 10)
Parameters:
 SVM-Type:  eps-regression
 SVM-Kernel:  linear
 cost:  10
 gamma:  0.01
 epsilon:  0.1
Number of Support Vectors:  34

print(svmmodel$index)
# these are support vectors:
  [1]  2  3  5  7  8  9 14 15 17 19 20 21 22 24 27 29 31 32 33 34 35 36 38 39 41 42 43 46 48
 54 55 60 61 64

print(svmmodel$coefs)
# these are coefficients of the support vectors:
svmmodel$coefs"

 [1,]  0.0020690035
 [2,] -0.0002287535
 [3,]  0.0007703764
 [4,] -0.0024933940
 [5,] -0.0029573383
 ...
[30,] -0.0013300716
[31,] -0.0020939429
[32,]  0.0022405260
[33,] -0.0044329149
[34,] -0.0007355469
```

The program also provides fitted values and decision numbers, which can be used to further examine the goodness of fit. We also use radial basis kernel to see the effect of classification to compare with the linear kernel used above.

```
# use svm in radial basis kernel

#Train the data with ranked frature with radial basis kernel
svmmodel <- svm(x[ , featureRankedList[1:100]], y, cost = 10, kernel="radial")
print(svmmodel)
# output:
Call: svm.default(x = x[, featureRankedList[1:100]], y = y, kernel = "radial", cost = 10)
Parameters:
 SVM-Type:  eps-regression
 SVM-Kernel:  radial
 cost:  10
 gamma:  0.01
 epsilon:  0.1
Number of Support Vectors:  44

print(svmmodel$index)
# these are support vectors:
  [1]  2  5  6  8  9 10 13 15 16 17 18 19 20 21 22 23 26 27 31 32 33 34 35 36 37 38 39 40
 [30] 41 43 44 45 47 48 49 50 52 53 54 55 56 60 62 63

print(svmmodel$coefs)
# these are coefficients of the support vectors:
svmmodel$coefs"

 [1,] -0.23798366
 [2,] -0.36076861
 [3,]  0.15172304
 [4,] -0.32956350
 [5,] -0.10064833
 ...
[40,] -0.71284955
```

```
[41,] -0.59391408
[42,] -0.16696255
[43,]  0.01255581
[44,] -0.20662406
```

One can notice that radial basis kernel selected more support vectors than the linear kernel, which provides better goodness of fit by using radial basis kernel. Therefore, one need to be aware the kernel used in the model fitting. The next section offers a more contrast comparison.

5.2.6   Example: Effect of Kernel Selection

The results of classification can be greatly affected by the selection of the kernel functions. In the monograph by Hastie et al. [32], the authors simulated a data set for two groups and use SVM to demonstrate the group identification by using a kernel function. The simulated data are analyzed again using both linear and radial basis functions and the graphical outputs are shown in Figure 5.2. The difference can be easily observed. However, in the analysis of real data with multivariate covariates, such graphs with clear cut classification boundary are seldom obtainable.

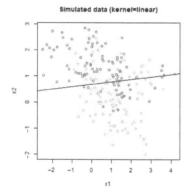 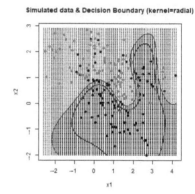

FIGURE 5.2
Classification via SVM with linear and radial basis kernels.

5.3   SVM for Continuous Data

When data set $\{y_i, i = 1, \cdots, n\}$ is normally distributed to evaluate the goodness of fit of a model $\{f(x, \theta), \theta \in A\}$, a common measure is to use the quadratic loss function

$$M(y, f(x, \theta)) = (y - f(x, \theta))^2.$$

However, if the error is generated by other distributions, different measure other than squared loss can be more efficient. For example, in the context of robust statistics, Huber (1964) had developed a theory on M-estimate that provided a strategy for choosing the loss function using general information of the error distribution.

To construct a SVM for real-valued functions with the possibility of non-normal data and to mimic the idea of margin in SVM for indicator function, the so-called ϵ-insensitive loss functions:

$$M(y, f(x, \theta)) = L(|y - f(x, \theta)|_\epsilon)$$

is used for the measure of loss function with

$$|y - f(x, \theta)|_\epsilon = \begin{cases} 0 & \text{if } |y - f(x, \theta)| \leq \epsilon; \\ |y - f(x, \theta)| - \epsilon & \text{otherwise.} \end{cases}$$

These loss functions describe the ϵ-insensitive model. The loss is assumed to be equal to 0 if the discrepancy between the predicted and the observed values is less than ϵ; otherwise, it will be assumed to be the value of the residuals minus ϵ.

5.3.1 Minimizing the Risk with ϵ-insensitive Loss Functions

Consider the general regression function $f(x) = \sum_{i=1}^{n} \psi_i x_i + b$ with $\Psi = (\psi_1, \cdots, \psi_n)$ and $X = (x_1, \cdots, x_n)$. Given a sample data set

$$\{(y_1, x_1), \cdots, (y_n, x_n)\},$$

our goal is to find the parameters Ψ and b that minimize the empirical risk

$$R_{emp}(\Psi, b) = (1/l) \sum_{i=1}^{l} |y_i - (\psi * x_i) - b|_{\epsilon_i},$$

under the constraint that vector of parameters Ψ satisfies

$$(\Psi * \Psi) \leq A^2, \tag{5.32}$$

and obtain a solution in the form of (5.21) satisfying certain constraints.

This optimization problem is equivalent to the problem of finding the pair (Ψ, b) that minimizes the quantity defined by slack variables $\xi_i, \xi_i^*, i = 1, \cdots, l$ under constraints

$$y_i - (\psi_i * x_i) - b \leq \epsilon + \xi_i^*,$$

$$-y_i + (\psi_i * x_i) + b \leq \epsilon + \xi_i,$$

$$\xi_i^* \geq 0 \quad \text{and} \quad \xi_i \geq 0$$

for $i = 1, \cdots, l$ and constraint (5.32).

To solve the optimization problem with constraints of inequality type, one has to find the saddle point of the Lagrange functional

$$L(\Psi, \xi^*, \xi, \alpha^*, \alpha, \gamma, \beta, \beta^*)$$

$$= \sum_{i=1}^{l}(\xi^* + \xi) - \sum_{i=1}^{l}\alpha_i[y_i - (\psi_i * x_i) - b + \epsilon_i + \xi_i]$$

$$- \sum_{i=1}^{l}\alpha_i^*[(\psi_i * x_i) + b - y_i + \epsilon_i + \xi_i^*] - (\gamma/2)(A^2 - (\Psi * \Psi))$$

$$- \sum_{i=1}^{l}(\beta_i^*\xi_i^* + \beta_i\xi_i). \tag{5.33}$$

The minimum is obtained with respect to elements ψ_i, b, ξ_i, and ξ_i^* and the maximum with respect to Lagrange multipliers $\gamma \geq 0, \alpha_i^* \geq 0, \alpha_i \geq 0, \beta_i^* \geq 0$, and $\beta_i \geq 0, i = 1, \cdots, l$.

Minimization with respect to Ψ, b, ξ_i, and ξ_i^* yields the following equalities:

$$\Psi = \sum_{i=1}^{l}\{(\alpha_i^* - \alpha_i)/\gamma\}x_i,$$

$$\sum_{i=1}^{l}\alpha_i^* = \sum_{i=1}^{l}\alpha_i,$$

$$\beta_i + \alpha_i^* \leq 1 \quad \text{and} \quad \beta_i + \alpha_i \leq 1,$$

for $i = 1, \cdots, l$. Substituting these equalities into (5.33), one obtains the following functional, which has to be further maximized with respect to the parameters mentioned above:

$$\tilde{L}(\alpha, \alpha^*, \gamma) = -\sum_{i=1}^{l}\epsilon_i(\alpha_i^* + \alpha_i) + \sum_{i=1}^{l}y_i(\alpha_i^* - \alpha_i)$$

$$- \frac{1}{2\gamma}\sum_{i,j=1}^{l}(\alpha_i^* - \alpha_i)(\alpha_j^* - \alpha_j)(x_i * x_j) - \frac{A^2\gamma}{2}. \tag{5.34}$$

By taking derivative of Equation (5.34) with respect to γ, one can obtain the estimate of γ as

$$\gamma = \left\{\sum_{i,j=1}^{l}(\alpha_i^* - \alpha_i)(\alpha_j^* - \alpha_j)(x_i * x_j)\right\}^{1/2}\Big/A, \tag{5.35}$$

Substitute equation (5.35) into (5.33), one has

$$\tilde{L}(\alpha, \alpha^*) = -\sum_{i=1}^{l} \epsilon_i(\alpha_i^* + \alpha_i) + \sum_{i-1}^{l} y_i(\alpha_i^* - \alpha_i)$$

$$- A\Big(\sum_{i,j=1}^{l} (\alpha_i^* - \alpha_i)(\alpha_j^* - \alpha_j)(x_i * x_j) \Big)^{1/2}. \quad (5.36)$$

The parameter b can be estimated by minimizing the empirical risk functional $R_{emp}(\Psi, b)$.

5.3.2 Example: Regression Analysis Using SVM

In this section, we illustrate the regression using SVM to identify the support vectors the other information. The large data of riboavin production with *Bacillus subtilis* (see http://www.annualreviews.org) were used.

```
library(e1071)
library(kernlab)

## prepare data (x,y) from original data
ribof<-t(riboflavin[-1,-1])
x <- ribof[,-1]
y <- ribof[,1]

svrmodel<-svm(y~x, epsilon = 0.2, cost = 1, gamma =0.125)
print(svrmodel)
# outputs:
Call: svm(formula = y ~ x, epsilon = 0.2, cost = 1, gamma = 0.125)
Parameters:
 SVM-Type:  eps-regression
 SVM-Kernel:  radial
 cost:  1
 gamma:  0.125
 epsilon:  0.2
Number of Support Vectors:  57

print(svrmodel$index)
# svrmodel$index of support vectors:
 [1]  1  2  3  4  6  7  8  9 11 12 13 14 15 16 17 18 19 21 22 24 28 29 31 32 33 34 35 36 37
[30] 39 40 41 42 43 45 46 47 49 50 51 52 53 55 56 57 58 59 60 61 62 63 64 66 67 68 69 70

print(svrmodel$coef)
# coefficients of the support vectors:
 [1,]  1.00000000
 [2,]  1.00000000
 [3,]  1.00000000
 [4,]  1.00000000
 [5,]  0.48699402
 ...
[53,] -0.51627716
[54,] -0.12860593
[55,]  1.00000000
[56,] -1.00000000
[57,]  0.81381816
```

The residual plot for svm is shown in Figure 5.3. In addition, the distribution of values of decision function for rvm and svm is shown in Figure 5.4. It is clearly that **rvm** has a finer distribution hence produces a better fit. (Note: rvm is discussed in Section 5.6.)

```
plot(svrmodel$fitted, svrmodel$residuals, ylim=c(-0.6,1.2), xlab="", ylab="")
lines(lowess(svrmodel$fitted, svrmodel$residuals), col="red")
abline(h=0)
title(main="SVM residuals vs fitted", xlab="Fitted values (svm)", ylab="Residuals (svm)")
hist(svrmodel$decision.values, main="Decision values (svm)",xlab="", ylab="Decision values")
```

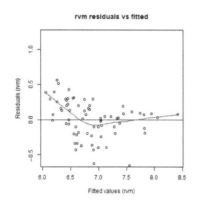
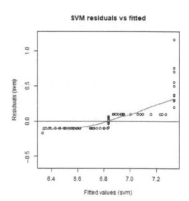

FIGURE 5.3

rvm and **svm** residuals vs fitted values plot.

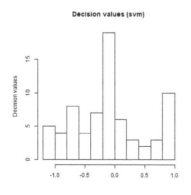
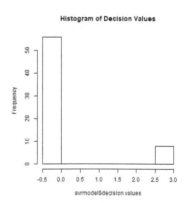

FIGURE 5.4

Decision value distribution of **rvm** (left) and **svm** (right) data analysis.

5.4　SVM for Survival Data Analysis

Research on applying SVM to survival data analysis has been ongoing for years, one issue of using support vector regression for censored data lies in the uncertainty about the outcomes. In order to use the information that censored data still provide partial information, Shivaswamy [77] proposed a support vector regression approach to censored data under the regular constraints for regression on continuous observations in addition to that for censored data. Khan [44] proposed penalty for loss in incorrect prediction of failure times for regression approaches.

Some researchers had framed survival problems as a classification problem instead of regression (e.g., [95], [15]) to address the question whether the patient survives a certain predefined landmark, e.g., 3-year or 5-year survival in many cancer studies. To obtain this goal, a SVM ranking method is used. The methods usually involve a regularization and a penalization for each comparable pair of data points for which the order in prediction differs from the observed order.

Van Belle et al. [92], [93], [94] proposed an approach based on both ranking and regression to analyze survival data using the ideas of transformation models (see e.g., Kalbfleisch and Prentice [41]). (Note: in the following, we will restrict the discussions in the data sample space instead of hyperplane of the feature space.)

Let $u(x) = w^t x + b$, based on Van Belle et al. [92], they estimated the prediction function by minimizing the empirical risk of mis-ranking two observations.

More formally, the goal is to minimize the error of mis-ranking subject to constraints:

$$\min_{w,\epsilon} \sum_{i=1}^{n} \epsilon_i \quad s.t. \quad \begin{cases} u(x_i) - u(x_{i-1}) + \epsilon_i - \epsilon_{i-1} \geq y_i - y_{i-1} \\ \epsilon_i - \epsilon_{i-1} \geq 0 \end{cases} \tag{5.37}$$

for $i = 1, \cdots, n$.

In addition to estimate the prediction function by minimizing the empirical risk on misrankings only, the authors also include regression constraints in SVM so that the model can be more flexible to incorporate the regression frameworks. The estimation hence can be obtained via the following constrained minimization with Lagrange multipliers as described in the previous sections.

$$\min_{w,\epsilon,\xi_i,\xi_i^*} (1/2)w^t w + \gamma \sum_{i=1}^{n} \epsilon_{i,i-1} + \mu \sum_{i=1}^{n} (\xi_i + \xi_i^*) \tag{5.38}$$

such that

$$
\begin{cases}
w^t(x_i - x_{i-1}) + \epsilon_{i,i-1} \geq (y_i - y_{i-1}), \\
w^t x_i + \xi_i \geq y_i, \\
-\delta_i w^t x_i + \xi_i^* \geq -\delta_i y_i, \\
\epsilon_{i,i-1} \geq 0, \ \epsilon_i \geq 0, \ \xi_i^* \geq 0, w_q \geq 0,
\end{cases}
\tag{5.39}
$$

for $i = 1, \cdots, n$, $q = 1, \cdots, p$ and $\epsilon_{i,i-1} = \epsilon_i - \epsilon_{i-1}$.

5.4.1   Example: Analysis of Survival Data Using SVM

In this section, we illustrate the procedures in the following using a large cancer data set from TCGA. The first 50 cases (tumor samples) were used with more than 16500 genes (features/variables). We tried to select the support vectors but not with feature selection, which will be discussed in Section 5.5.

```
library(survivalsvm)

## this is the analysis using type = regression:
survsvmREG <- survivalsvm(formula = Surv(AVAL, CNSR) ~ ., data = xdat, type = "regression",
            gamma.mu = 0.1, opt.meth = "quadprog", diff.meth = "makediff3", kernel = "lin_kernel",
            sgf.sv = 5, sigf = 7, maxiter = 20, margin = 0.05, bound = 10)

print(survsvmREG)
# outputs: describe the results
Call: survivalsvm(formula = Surv(AVAL, CNSR) ~ ., data = xdat, type = "regression", gamma.mu = 0.1,
            opt.meth = "quadprog", diff.meth = "makediff3", kernel = "lin_kernel", sgf.sv = 5,
            sigf = 7, maxiter = 20, margin = 0.05, bound = 10)
Survival svm approach                 : regression
Type of Kernel                        : lin_kernel
Optimization solver used              : quadprog
Number of support vectors retained : 5
survivalsvm version                   : 0.0.5

print(survsvmREG$model.fit$Beta)
# output: these are the support vectors with non-zero coefs.
[1] -0.00018  0.00015  0.00006  0.00020 -0.00023


## this is the analysis using type = vanbelle1:
survsvmREG <- survivalsvm(formula = Surv(AVAL, CNSR) ~ ., data = xdat, type = "vanbelle1",
            gamma.mu = 0.1, opt.meth = "quadprog", diff.meth = "makediff3", kernel = "lin_kernel",
            sgf.sv = 5, sigf = 7, maxiter = 20, margin = 0.05, bound = 10)
print(survsvmREG)
# outputs: describe the results
Call: survivalsvm(formula = Surv(AVAL, CNSR) ~ ., data = xdat, type = "vanbelle1", gamma.mu = 0.1,
            opt.meth = "quadprog", diff.meth = "makediff3", kernel = "lin_kernel", sgf.sv = 5, sigf = 7,
            maxiter = 20, margin = 0.05, bound = 10)
Survival svm approach                     : vanbelle1
Type of Kernel                            : lin_kernel
Method used to build 1NN differences    : diff3
Optimization solver used                  : quadprog
Number of support vectors retained      : 25
survivalsvm version                       : 0.0.5

print((survsvmREG$model.fit)$Alpha)
# output: support vectors have non-zero coefs.
 [1] 0.00000 0.00000 0.00000 0.00094 0.00233 0.00000 0.00316 0.00000 0.00000 0.00000 0.00000 0.00000
[13] 0.00000 0.00355 0.00000 0.00000 0.00000 0.00000 0.00000 0.00000 0.00000 0.00000 0.00000 0.00000
[25] 0.00550 0.00650 0.00000 0.00716 0.00721 0.00007 0.00000 0.00000 0.00782 0.00827 0.00856 0.00834
[37] 0.00000 0.00860 0.00863 0.00793 0.00068 0.00629 0.00571 0.00491 0.00399 0.00392 0.00288 0.00178 0.00095
```

```
## this is the analysis using type = vanbelle2:
survsvmREG <- survivalsvm(formula = Surv(AVAL, CNSR) ~ ., data = xdat, type = "vanbelle2",
                gamma.mu = 0.1, opt.meth = "quadprog", diff.meth = "makediff3",
                kernel = "lin_kernel", sgf.sv = 5, sigf = 7, maxiter = 20, margin = 0.05,
                bound = 10)
print(survsvmREG)
# outputs: describe the results
Call: survivalsvm(formula = Surv(AVAL, CNSR) ~ ., data = xdat, type = "vanbelle2", gamma.mu = 0.1,
                opt.meth = "quadprog", diff.meth = "makediff3", kernel = "lin_kernel", sgf.sv = 5,
                sigf = 7, maxiter = 20, margin = 0.05, bound = 10)
Survival svm approach               : vanbelle2
Type of Kernel                      : lin_kernel
Method used to build 1NN differences : diff3
Optimization solver used            : quadprog
Number of support vectors retained  : 27
survivalsvm version                 : 0.0.5


print((survsvmREG$model.fit)$Alpha)
# output: support vectors have non-zero coefs.
 [1] 0.00000 0.00000 0.00000 0.00311 0.00730 0.00000 0.01028 0.00000 0.00000 0.00000 0.00000 0.00000
[13] 0.00000 0.01217 0.00000 0.00000 0.00000 0.00000 0.00000 0.00000 0.00000 0.00000 0.00000 0.00000
[25] 0.01778 0.01858 0.00000 0.01910 0.01862 0.00125 0.00005 0.00000 0.01766 0.01791 0.01791 0.01702
[37] 0.00049 0.01678 0.01639 0.01482 0.00195 0.01090 0.01025 0.00914 0.00732 0.00698 0.00577 0.00429
0.00206
```

One can notice that the results are difference using different **types**, which describe the different objective functions to be minimized with the specific set of constraints. Using **regression** type resulted with 5 support vectors, but using **vanbelle1** or **vanbelle2** resulted with 25 and 27 support vectors, respectively. In addition, the support vectors are not always identical, which should not be a surprise with big data. Generally, the **vanbelle2** type is preferred because it takes into account of the regression errors and the concordance of prediction error of censored cases. Analysts need to be aware of the differences in using this kind of methodology.

5.5 Feature Elimination for SVM

One common approach to analyze data with a large number of variables (covariates) is to conduct a dimension reduction procedure to reduce the number of covariates. For example, one of these scenarios is the gene selection in genomics research as the data usually have a large number of genes. Many algorithms have been developed in the literature, e.g., using ranking criteria with correlation or hypothesis testing. Many of these methods select important variables individually and fail to consider the mutual information such as collinearity or interactions among variables.

Standard SVM does not usually conduct dimension reduction by eliminating the variables; therefore, the methods for variable reduction (aka feature elimination) had been proposed by many researchers in the past decades (see Zhao et al. [103]). Zhao et al. [103] proposed another method to deal with this issue and their procedure has been implemented in R package `sigFeature`. They formulate the SVM as a regularization problem with penalty function SCAD proposed by Fan and Li [19] instead of the usual L_1.

Specifically, given a training set $\{(x_i, y_i), i = 1, \cdots, n \; x \in R^p, y \in \{0, 1\}\}$, and the classification rule $f(\cdot)$. As discussed in the previous sections, one can assume that f has the following general form

$$f(x, w) = \phi(x) \times w + b,$$

Zhao et al. proposed to minimize

$$\min_{b,w}(1/n) \sum_{i=1}^{n} \{1 - y_i f(x, w)\}_+ + \sum_{j=1}^{p} \mathcal{S}_\lambda(|w_j|), \tag{5.40}$$

where $\mathcal{S}$ is the SCAD function.

(Note: even though their paper was in the setting of gene selection; however, the procedures can be applied to any large data with many variables.)

5.5.1 Example: Gene Selection via SVM with Feature Elimination

In the following, we illustrate the implementation of the procedures using a large gene set. This example is taken from the example in Das et al [14].

```
library(sigFeature)
library(SummarizedExperiment)

data(ExampleRawData, package="sigFeature")
x <- t(assays(ExampleRawData)$counts)  # gene expression data
y <- colData(ExampleRawData)$sampleLabels  # sample label (1 or -1)

## this part is for data analysis using sigFeature method:
pvals <- sigFeaturePvalue(x,y)
print("The p-values:")
print(unlist(pvals)[1:10])
 [1] 0.022449062 0.040997843 0.046433665 0.040765741 0.007009415
     0.024470011 0.042995820 0.039783798 0.017253246 0.019366123

data(sigfeatureRankedList)
print("Top 10 most significant genes from sigFeature:")
print(colnames(x)[sigfeatureRankedList[1:10]])
 [1] "221596_s_at" "204250_s_at" "221292_at"   "221310_at"   "216130_at"
     "216704_at"   "215542_at"   "222038_s_at" "211201_at"   "206811_at"

sigFeature.model=svm(x[,sigfeatureRankedList[1:1000]], y, type="C-classification",
                kernel="linear")
print(summary(sigFeature.model))

# output:
Call: svm.default(x = x[, sigfeatureRankedList[1:1000]], y = y, type = "C-classification",
                kernel = "linear")
Parameters:
   SVM-Type:  C-classification
   SVM-Kernel:  linear
   cost:  1
Number of Support Vectors:  16 ( 10 6 )
```

```
Number of Classes:  2 (-1 1)

rownames(sigFeature.model$coefs)<-rownames(x)[sigFeature.model$index]
print(sigFeature.model$coefs)

# output:
# this is the 16 support vector genes with the coefficients.
# the first 10 with positive values are in one category, the other 6 are in the second
  category.

support    coefs
vector
GSM42250   6.531854e-04
GSM42252   4.281329e-04
GSM42254   3.213426e-05
GSM42249   3.449226e-04
GSM42251   8.210336e-04
GSM42255   6.731978e-04
GSM42260   5.237181e-04
GSM42269   6.973117e-04
GSM42270   1.477157e-04
GSM42271   7.383887e-05
GSM42262  -5.872142e-05
GSM42264  -3.813943e-04
GSM42265  -3.769902e-04
GSM42266  -1.502197e-03
GSM42267  -7.330802e-04
GSM42268  -1.342808e-03

## this part is for data analysis using SVM-RFE method:
data(featureRankedList)
print("Top 10 features are printed below:")
print(colnames(x)[featureRankedList[1:10]])
  [1] "211943_x_at" "215157_x_at" "212788_x_at" "200064_at"   "213699_s_at"
      "216520_s_at" "201631_s_at" "213275_x_at" "208956_x_at" "201201_at"

RFE.model=svm(x[ ,featureRankedList[1:1000]], y, type="C-classification", kernel="linear")
print(summary(RFE.model))

# output:
Call: svm.default(x = x[, featureRankedList[1:1000]], y = y, type = "C-classification",
                  kernel = "linear")
Parameters:
   SVM-Type:  C-classification
   SVM-Kernel:  linear
   cost:  1
Number of Support Vectors:  17 ( 10 7 )
Number of Classes:  2 ( -1 1)

rownames(RFE.model$coefs)<-rownames(x)[RFE.model$index]
print(RFE.model$coefs)

# output:
# this is the 17 support vector genes with the coefficients.
# the first 10 with positive values are in one category, the other 7 are in the second
  category.

support    coefs
vector
GSM42250   5.829137e-04
GSM42252   2.801012e-04
GSM42254   2.002726e-04
GSM42249   1.296759e-04
GSM42251   4.294254e-04
GSM42255   9.246166e-04
GSM42260   4.848147e-04
GSM42269   1.014662e-03
```

```
GSM42270   1.644586e-04
GSM42271   9.147396e-05
GSM42263  -3.301780e-04
GSM42264  -3.299134e-05
GSM42265  -1.142249e-03
GSM42266  -3.787036e-04
GSM42267  -6.483022e-04
GSM42268  -1.445051e-03
GSM42272  -3.249399e-04
```

Comparing the genes selected by these two methods and the significance of the genes, one can notice the differences. The top few genes from these two methods are not the same and their significance is also quite different. In the following, we compare the significance of the genes selected by sigFeature and SVM-RFE methods. There are some substantial difference between the genes selected and the p-values of these genes in prediction of outcomes. Figure 5.5 shows the p-values of the genes from sugFeature and SVM-RFE, the contrast is easy to compare. Figure 5.6 shows the box plot of the p-values of the genes from sigFeature and SVM-RFE, again, the contrast is easy to observe.

```
pvalsigFe <- sigFeaturePvalue(x, y, 100, sigfeatureRankedList)
pvalRFE <- sigFeaturePvalue(x, y, 100, featureRankedList)
#par(mfrow=c(1,2))
hist(unlist(pvalsigFe),breaks=50, col="skyblue", main=paste("P-values by sigFeature"),
    xlab="The top 100 most significant p values")
box()
hist(unlist(pvalRFE),breaks=50, col="skyblue", main=paste("P-values by SVM-RFE"),
    xlab="The top 100 most significant p values")
box()

mytitle<-'P-values of sig. genes (sigFeature & SVM-RFE)'
boxplot(unlist(pvalsigFe), unlist(pvalRFE), main=mytitle, names=c("sigFeature", "SVM-RFE"),
        ylab="p value", ylim=c(min(unlist(pvalsigFe)), max(unlist(pvalRFE))))
stripchart(unlist(pvalsigFe), vertical=TRUE, method="jitter", add=TRUE, pch=16, col=c('green'))
stripchart(unlist(pvalRFE), vertical=TRUE, at=2, method="jitter", add=TRUE, pch=16, col=c('blue'))
grid(nx=NULL, ny=NULL, col="black", lty="dotted")
```

5.6   Spare Bayesian Learning with Relevance Vector Machine (RVM)

The discussions above are all about deterministic results either in regressions or classifications. Tipping [87] proposed another approach to use SVM for data analysis. The rationales are (1) the regular SVM makes liberal use of basis functions since the number of support vectors required typically grows linearly with the size of the training set, (2) the predictions are not probabilistic, (3) SVM needs to estimate the error/margin trade-off parameter for classification and the insensitivity parameter ϵ for regression, and (4) the kernel function must satisfy Mercer's condition if kernels are used.

RVM is a Bayesian counterpart of SVM and does not suffer from any of the above limitations. It adopts a fully probabilistic framework and the most

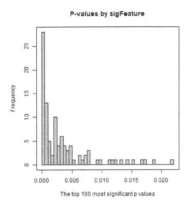
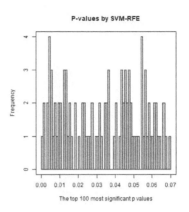

FIGURE 5.5
p-values of the genes from sigFeature and SVM-RFE.

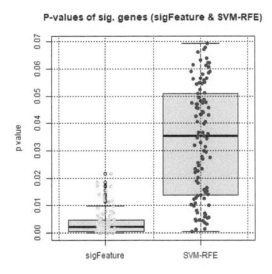

FIGURE 5.6
p-values of the genes from sigFeature and SVM-RFE.

probable values of the assumed priors over the model weights (parameters) are iteratively estimated from the data. Model sparsity is achieved because in practice the posterior distributions of many of the weights peak around zero. The training vectors with non zero weights are called relevance vectors.

Mathematically, consider the general regression setting with the response $y(x)$:

$$y(\mathbf{x}; w) = \sum_{i=1}^{M} w_i \phi_i(x) = \mathbf{w}^T \Phi(x) \tag{5.41}$$

where $\Phi(x) = (\phi_1(x), \phi_2(x), \cdots, \phi_M(x))^T$ are basis functions. As seen in the previous sections, the general SVM equation (5.41) can be expressed with kernel functions as

$$y(\mathbf{x}; w) = \sum_{i=1}^{M} w_i K(\mathbf{x}, x_i) + w_0. \tag{5.42}$$

RVM considers the following model: for an input data $\{\mathbf{x}_i, t_i\}, i = 1, \cdots, N$,

$$t_i = y(\mathbf{x}_i, w) + \epsilon_i,$$

with $\epsilon \sim N(\cdot, \sigma^2)$, hence the probability of t_i is $p(t_i|\mathbf{x}) = N(t_i|y(x_n), \sigma^2)$. The likelihood function of all data can be expressed as

$$p(\mathbf{t}|\mathbf{w}, \sigma^2) = (2\pi\sigma^2)^{-N/2} \exp\left\{ -1/(2\sigma^2) \times ||\mathbf{t} - \Phi\mathbf{w}||^2 \right\}.$$

To avoid over-fitting, the model imposes some Bayesian constraints including

$$P(w|\alpha) = \prod_{i=0}^{N} N(w_i|0, \alpha_i^{-1}),$$

$$p(\alpha) = \prod_{i=0}^{N} \text{Gamma}(\alpha_i|a, b), \quad p(\beta) = \text{Gamma}(\beta|c, d)$$

where $\beta = \sigma^{-2}$ and $\text{Gamma}(\alpha|a, b) = \Gamma(a)^{-1} b^a \alpha^{a-1} e^{-b\alpha}$.

To make prior non-informative, one can set a, b, c, d to a very small number, e.g., 10^{-4}, and use the following Bayes' rule for estimation and inference:

$$p(\mathbf{w}, \alpha, \sigma^2|\mathbf{t}) = p(\mathbf{t}|\mathbf{w}, \alpha, \sigma^2) p(\mathbf{w}, \alpha, \sigma^2)/p(\mathbf{t}).$$

The posterior distribution of the weights can be estimated using

$$p(\mathbf{w}|\mathbf{t}, \alpha, \sigma^2) = p(\mathbf{t}|\mathbf{w}, \sigma^2) p(\mathbf{w}|\alpha)/p(\mathbf{t}|\alpha, \sigma^2) \tag{5.43}$$

$$= (2\pi)^{-(N+1)/2} |\Sigma|^{-1/2} \exp\left\{ -\frac{1}{2}(w - \mu)^t \Sigma^{-1}(w - \mu) \right\}$$

with the posterior mean and covariance matrix as

$$\mu = \sigma^{-2} \Sigma \Phi^t t, \quad \Sigma = (\sigma^{-2} \Phi^t \Phi + A)^{-1},$$

where $A = \text{diag}(\alpha_0, \alpha_1, \cdots, \alpha_N)$.

5.6.1   Example: Regression Analysis Using RVM

In this section, we illustrate the regression using RVM to identify the support vectors the other information. The large data of riboavin production with *Bacillus subtilis* (see http://www.annualreviews.org) were used.

```
library(e1071)
library(kernlab)

## prepare data (x,y) from original data
ribof<-t(riboflavin[-1,-1])
x <- ribof[,-1]
y <- ribof[,1]

fitsvr<-svrmodel<-rvm(y ~ x)
print(svrmodel)
# outputs:
Relevance Vector Machine object of class "rvm"
Problem type: regression
Gaussian Radial Basis kernel function.
Hyperparameter : sigma =  0.000810796075162476
Number of Relevance Vectors : 30
Variance :  0.1012977
Training error :  0.065292253

print(RVindex(fitsvr))
# print relevance vectors
 [1]  1  2  3  5  7 10 12 14 16 17 20 23 26 30 33 35 36 44 46 49 50 51 58 61 63 65 67 68 70 71

print(alpha(fitsvr))
[1] "# the alpha values"
 [1]  3.4135619 -1.8743427  4.1362365 -2.5803577  3.3490531  1.2670073 -1.6692115 -0.5313602
 [9]  3.0097055  1.5452908 -3.0036526  2.6791590  2.5111235 -0.6177715  0.4058407  0.7541557
[17]  1.9022899  1.8767164  3.3784408 -0.9002568  3.3654645  4.6967180  1.1241537  3.7147848
[25] -0.4244736  2.0481612 -3.2451234  1.5946901  2.8382203  1.8348861
```

The fitted values plotted against the residuals are shown in Figure 5.4 using the following codes. The `lowess` curve shows certain curvature, which indicates a higher degree terms of the covariate that may be needed to improve the overall data fitting.

```
plot(fitted(fitsvr), ymatrix(fitsvr)-fitted(fitsvr), ylim=c(-0.6,1.2), xlab="", ylab="")
title(main="rvm residuals vs fitted", xlab="Fitted values (rvm)", ylab="Residuals (rvm)")
lines(lowess(fitted(fitsvr), ymatrix(fitsvr)-fitted(fitsvr)), col="red")
abline(h=0)
```

5.6.2   Example: Curve Fitting for SVM and RVM

In this section, we show the difference of goodness of fit using SVM and RVM. A data set was simulated with $\sin(x)/x$ for $x = \text{seq}(-10, 10, 0.1) + \text{rnorm}(0, 0.03)$.

```
library(e1071)
library(kernlab)

## simulate data
x <- seq(-10,10,0.1)
y <- sin(x)/x + rnorm(201,sd=0.03)
```

```
plot(x,sin(x)/x, type="n", main="Original simulated data", xlab="x values",
                            lab="y values: sin(x)/x")
points(x,sin(x)/x, pch="*",cex=1)
regm <- svm(x,sin(x)/x,epsilon=0.01,kpar=list(sigma=16),cross=3)
print(regm)
# output:
Call: svm.default(x = x, y = sin(x)/x, epsilon = 0.01, cross = 3, kpar = list(sigma = 16))
Parameters:
 SVM-Type:  eps-regression
 SVM-Kernel:  radial
 cost:  1
 gamma:  1
 epsilon:  0.01
Number of Support Vectors:  200

lines(x,predict(regm,x),col="red", lty=2)

regm <- rvm(x,sin(x)/x,epsilon=0.01,kpar=list(sigma=16),cross=3)
print(regm)
#output
Relevance Vector Machine object of class "rvm"
Problem type: regression
Gaussian Radial Basis kernel function.
 Hyperparameter : sigma =  16
Number of Relevance Vectors : 126
Variance :  2e-09
Training error : 1e-09
Cross validation error : 0.001513841

lines(x,predict(regm,x),col="black", lty=1)


plot(x,y, type="n", main="Original simulated data plus error", xlab="x values",
                            ylab="y values: sin(x)/x + error ")
points(x,y, pch="*",cex=1)
regm <- svm(x,y,epsilon=0.01,kpar=list(sigma=16),cross=3)
print(regm)
# output:
Call: svm.default(x = x, y = y, epsilon = 0.01, cross = 3, kpar = list(sigma = 16))
Parameters:
 SVM-Type:  eps-regression
 SVM-Kernel:  radial
 cost:  1
 gamma:  1
 epsilon:  0.01
Number of Support Vectors:  195

lines(x,predict(regm,x),col="red", lty=2)

regm <- rvm(x,y,epsilon=0.01,kpar=list(sigma=16),cross=3)
print(regm)
#output:
Relevance Vector Machine object of class "rvm"
Problem type: regression
Gaussian Radial Basis kernel function.
 Hyperparameter : sigma =  16
Number of Relevance Vectors : 57
Variance :  0.000709815
Training error : 0.000533719
Cross validation error : 0.009615746

lines(x,predict(regm,x),col="black", lty=1)
```

One can easily notice that SVM generally uses more supporting vectors than RVM in either fitting the original or random error added data. The curve fitting of the predicted values is shown in Figure 5.7. The data points are shown in '*', the fitted values by SVM are shown in dash line, and the fitted values by RVM are shown in solid line. The fit by RVM generally tracks the data points much closer than that from SVM.

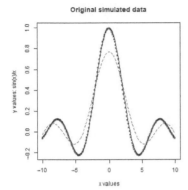
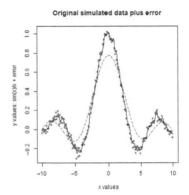

FIGURE 5.7
Curve fitting with SVM (dash line) and RVM (solid line).

5.7 SV Machines for Function Estimation

To construct the support vector machine for real-valued function estimation problems, we map the input vectors x into high-dimensional feature space Z, where we consider linear functions

$$f(x, \beta) = (z * \Psi) + b = \sum_{i=1}^{l} \beta_i (z * z_i) + b.$$

This function expressed in feature space can also be represented in input space as:

$$f(x, \beta) = (z * \Psi) + b = \sum_{i=1}^{l} \beta_i K(x, x_i) + b, \qquad (5.44)$$

where for $i = 1, \cdots, l$, β_i are scalars; x_i are vectors; and $K(x, x_i)$ is a given function satisfying Mercer's conditions.

To minimize

$$R_{emp}(\Psi, b) = \frac{1}{l} \sum_{i=1}^{l} |y - (\psi_i * z_i) - b|_{\epsilon_i}$$

subject to constraint as in the sample space, one has to maximize the following functionals with respect to $\alpha_i^*, \alpha_i, \gamma$

$$W(\alpha, \alpha^*, \gamma) = -\sum_{i=1}^{l} \epsilon_i(\alpha_i^* + \alpha_i) + \sum_{i=1}^{l} y_i(\alpha_i^* - \alpha_i)$$

$$- A \times \left\{ \sum_{i,j=1}^{l} (\alpha_i^* - \alpha_i)(\alpha_j^* - \alpha_j) K(x_i, x_j) \right\}^{1/2}. \qquad (5.45)$$

subject to constraints

$$\sum_{i=1}^{l} \alpha_i^* = \sum_{i=1}^{l} \alpha_i$$

and

$$0 \leq \alpha_i^*, \alpha_i \leq 1, \quad i = 1, \cdots, l.$$

For the kernel $K(x, x_i)$, the polynomial kernel:

$$K(x, x_i) = |(x * x_i) + 1|^d.$$

and the radial basis kernel function

$$K(x, x_i) = K(|(x - x_i)|)$$

for example,

$$K(|(x - x_i)|) = \exp\{-\gamma |x - x_i|^2\}.$$

are among the most commonly used.

6

Cluster Analysis

As stated in Chapter 1, how to deal with data heterogeneity is one of the upmost tasks in data analysis. In big data, it is inevitable that data will come from several populations, hence data heterogeneity will naturally happen. Therefore, any data analyst should not expect one model will be sufficient to interpret the relationship among the variables. Hence, how to subgroup data and make inference for each of them becomes an important task at the outset of data analysis. The supervised tree-based modeling, which we had covered in Chapter 4, is one of the methods for his purpose. In this chapter, we will discuss data sub grouping from an unsupervised approach, namely, to classify data into clusters even though the data points do not have a assigned label or response variable.

Clustering is a common unsupervised machine learning task, in which the data set has to be automatically partitioned into clusters, such that objects within the same cluster are more similar, while objects in different clusters are more different. Unfortunately, there is not a generally accepted definition of a cluster; therefore, over many years of research and many clustering algorithms and evaluation measures have been proposed, each with their merits and drawbacks. Nevertheless, a few methods such as hierarchical clustering, K-means, and PAM (Kaufman and Rousseeuw [42],[43]) have received repeated and widespread use. In the following, we will describe these methods in addition to a bagged clustering method and how the `randomForest` can also be used for clustering.

6.1 Measure of Distance/Dissimilarity

Distance or similarity between data points is the foundation of any clustering strategy as the observations in the same cluster are supposed to be more similar than those in other clusters. Therefore, defining the measurements of distance or similarity is critical. Sometimes, even with the same data, multiple dissimilarities can be defined based on the objectives of the analysis. In this section, except for the well-known Euclidean distance, we describe a few commonly used measures for continuous, discrete, or even mixed data types of observations.

DOI: 10.1201/9781003205685-6

6.1.1   Continuous Variables

The Minkowski (L_q) Distance: the distance between two observations x_i, x_j of p-dimensional real-valued variables $x_i = (x_{i1}, \cdots, x_{ip})$ is defined as

$$d_{L_q}(x_i, x_j) = \left\{ \sum_{r=1}^{p} |x_{ir} - x_{jr}|^q \right\}^{1/q}.$$

By changing the value of q, one can obtain different kind of distance measures. For example, the Euclidean distance is d_{L_2} and the so-called Manhattan/city block distance is d_{L_1}. Note that the Minkowski distances are not scale equivariant. The value of d_{L_q} will be dominated by variables with larger variation and will change if variables are multiplied by constants. Therefore, data need to be properly transformed if needed to avoid this property of the measure.

The Mahalanobis Distance: The distance between two observations x_i, x_j of p-dimensional real-valued variables $x_i = (x_{i1}, \cdots, x_{ip})$ with sample covariance matrix $\hat{\Sigma}$ is defined as

$$d_M(x_i, x_j) = \left\{ (x_i - x_j)\hat{\Sigma}(x_i - x_j)' \right\}^{1/2}$$

6.1.2   Binary and Categorical Variables

If observation $\{x_i = (x_{i1}, \cdots, x_{ip}) | x_{ij} \in \{0,1\}, i = 1, \cdots, n\}$, namely, consisting of binary outcomes, then the most straightforward way to define a dissimilarity is to count the number of variables on which two objects x_1 and x_2 do not coincide, namely, the so-called Simple Matching Distance,' which is defined as

$$d_{SM}(x_1, x_2) = (1/p) \sum_{j=1}^{p} I(x_{1j} \neq x_{2j})$$

where $I(\cdot)$ is the indicator function.

Another measure, the "Jaccard distance," counts the number of entries indicating positive response, such as 1 in (0,1) outcomes. The Jaccard distance is defined as

$$d_J(x_1, x_2) = 1 - \frac{\sum_{j=1}^{p} I(x_{1j} = 1 \text{ and } x_{2j} = 1)}{\sum_{j=1}^{p} I(x_{1j} = 1 \text{ or } x_{2j} = 1)}.$$

When the variables have more categories than binary, one popular approach is to properly dichotomize the outcomes into two separate categories and then use the distance measures as defined above.

6.1.3   Mixed Data Types

Generally, data of large size will contain multiple types, including continuous and categorical (ordered or un-ordered) observations. For example, income, movie ratings, or gender, etc. In order to form cluster of data with these kind of data, a proper measure of distance or similarity needs to be defined.

Gower [28] proposed a general formula that can accommodate different dissimilarity measures.

Let $X_i = (x_{i1}, x_{i2}, \cdots, x_{iK})$ be the ith data point with K variables, then for each pair of data points $1 \leq i, j \leq N$ and variable k:

- Define similarity score $s_{i,j,k} = 0$ if x_{ik} and x_{jk} are dissimilar, and $0 < s_{i,j,k} \leq 1$, if x_{ik} and x_{jk} are not dissimilar depending on the degree of dissimilarity.

- Define $\delta_{i,j,k} = 0$ if x_{ik} and x_{jk} cannot be compared due to missing data, and $\delta_{i,j,k} = 1$ if x_{ik} and x_{jk} can be compared.

- Define the similarity score for X_i and X_j by

$$S_{ij} = \sum_{k=1}^{K} s_{i,j,k} / \sum_{k=1}^{K} \delta_{i,j,k}.$$

- If the variables are not equally important and certain weighting scheme $[w_k(x_{ik}, x_{jk})]$ is available, one can refine the equation above by incorporating the weights as

$$S_{ij} = \sum_{k=1}^{K} s_{i,j,k} w_k(x_{ik}, x_{jk}) / \sum_{k=1}^{K} \delta_{i,j,k} w_k(x_{ik}, x_{jk}).$$

6.1.4 Other Measure of Dissimilarity

Besides the measures of dissimilarity described above, several special purpose of measures had also been proposed. For example, especially in longitudinal data analysis, the patterns or the trends of responses over time can be of interest. In this case, the correlation of response patterns between observations can be estimated and the study subjects can be clustered according to the patterns.

In addition, the proximity measure of `randomForest` can also be used to measure dissimilarity between observations of mixed-types.

6.2 Hierarchical Clustering

Hierarchical clustering was proposed by Florek et al. [22] using the Single Linkage clustering in order to group objects based on a given dissimilarity. The aim of hierarchical clustering is to set up a hierarchy of clusters, i.e., not only to partition the data into a fixed number of clusters but rather to have a sequence of groupings at different levels, with member sets of finer groupings being subsets of coarser groupings.

Hierarchy clustering is usually presented graphical as an up-side-down (or side way) tree, the so-called dendrogram. Considering this as an up-side-down tree, the highest point of the tree represents the root-node which includes all the data observations. As the height is reduced, branches start to grow, which divide all the data at the root node into subsets. Again, as the heights are further reduced, smaller branches grow out of these bigger branches and the subsets of data are further divided into smaller subsets and be allocated into these smaller branches (or nodes). As this process repeats itself, one will arrive at the very tips of the tree with a single observation in each tip.

Practically, there are two main types of methods to create the hierarchies. First, the divide method which was just described above. Second, the agglomerative method, which is the reversal of the divide method. Namely, it starts from each observation and proceeds by merging with the closest (will be defined below) observation to form a new node. Continue this process until all the branches are merged together into the root node.

6.2.1   Options of Linkage

During the clustering process, each observation or cluster needs to find another observation or cluster which is closest (or most similar) so that they can be merged to form a new cluster. The following are a few most commonly used methods in practice to define the closeness.

Single Linkage (or nearest neighbor) clustering: For two clusters A and B, the distance is defined as

$$D_{SL}(A, B) = \min_{x \in A, y \in B} d(x, y)$$

Complete Linkage (or furthest neighbor) clustering: For two clusters A and B, the distance is defined as

$$D_{CL}(A, B) = \max_{x \in A, y \in B} d(x, y)$$

Average Linkage clustering: For two clusters A and B, the distance is defined as

$$D_{AL}(A, B) = \frac{1}{|A||B|} \sum_{x \in A, y \in B} d(x, y)$$

Examining these definitions, it is easy to note that Complete Linkage emphasizes within-cluster homogeneity but not the separation between clusters. In many cases, Complete Linkage clusters are not properly separated. On the other hand, Single Linkage emphasizes more on between-cluster separation but not as much on cluster homogeneity. Actually, Single Linkage clusters can be quite heterogeneous. However, Average Linkage does not have a straightforward interpretation as the other two linkage methods do. It is a compromise between Single and Complete Linkage, between within-cluster homogeneity and between cluster separation. Given these differences, it is always prudent for the analysts to try a set of different clustering methods and examine the findings for a proper meaningful interpretation.

Ward's minimum variance clustering: Ward's method is a criterion applied in hierarchical cluster analysis quite frequently. Ward's minimum variance method is a special case of the objective function approach originally presented by Ward [98]. This is an approach for performing cluster analysis by considering cluster analysis as an analysis of variance problem instead of using distance metrics or measures of association. Based on the optimal value of an objective function, Ward [98] suggested a general clustering procedure by choosing the pair of clusters to merge at each step. This objective function could be any function that reflects the purpose of the data analysis. Many of the standard clustering procedures are contained in this very general class.

Ward's minimum variance criterion minimizes the total within-cluster variance assuming Euclidean distance between data points. To implement this method, all clusters contain a single point at the initial step. At the subsequent steps, it finds the pair of clusters that lead to minimum increase in total within-cluster variance after merging. This increase is a weighted squared distance between cluster centers.

6.2.2 Example of Hierarchical Clustering

Figure 6.1 is a simulated data `mlbench.cassini` with three well-defined clusters indicated by the different colors. In this section, clusters will be produced using this data set with various linkages listed above. R-codes used are shown below as well as the respective outputs and the dendrograms are also shown below. As one can easily see that the different linkage method may create different cluster membership; therefore, it is prudent to explore various methods and compare the outputs to examine the differences and commonalities.

```
library(mlbench)
library(cluster)

cassini<-mlbench.cassini(1000)
cassini4clust<-cassini$x
plot(cassini4clust[,1], cassini4clust[,2], main="cassini clusters", xlab="", ylab="",
col=cassini$classes)

xx<-dist(cassini4clust, method = "euclidean")

hc<-hclust(xx, method = "single")
plot(hc, labels=FALSE, xlab="", main="Dendrogram (single)", sub="")
print(hc)

Call: hclust(d = xx, method = "single")
Cluster method   : single
Distance         : euclidean
Number of objects: 1000

cutk3<-cutree(hc, k=3) # cut into 3 clusters
```

cassini clusters

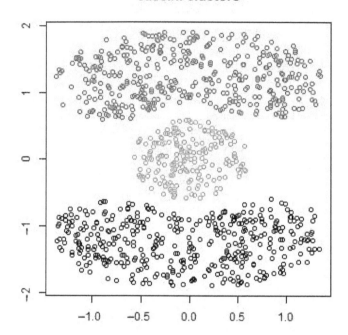

FIGURE 6.1
Simulated cassini data clusters.

```
print(cutk3) # this statement print the cluster membership for each case (omit print)
clustSize<-c(sum(cutk3==1),sum(cutk3==2),sum(cutk3==3))
print(clustSize) # print cluster size
[1] 400 400 200
[1] 400 399   1 200  # if cutree into 4 clusters

hc<-hclust(xx, method = "complete", members=NULL)
plot(hc, labels=FALSE, xlab="", main="Dendrogram (complete)", sub="")
cutk3<-cutree(hc, k=3)
clustSize<-c(sum(cutk3==1),sum(cutk3==2),sum(cutk3==3))
print(clustSize)
[1] 412 214 374
[1] 230 182 214 374  # if cutree into 4 clusters

hc<-hclust(xx, method = "average", members=NULL)
plot(hc, labels=FALSE, xlab="", main="Dendrogram (average)", sub="")
cutk3<-cutree(hc, k=3)
clustSize<-c(sum(cutk3==1),sum(cutk3==2),sum(cutk3==3))
```

```
clustSize<-c(sum(cutk4==1),sum(cutk4==2),sum(cutk4==3),sum(cutk4==4))
print(clustSize)
[1] 400 400 200
[1] 190 210 400 200 # if cutree into 4 clusters

hc<-hclust(xx, method = "ward", members=NULL)
plot(hc, labels=FALSE, xlab="", main="Dendrogram (ward)", sub="")
cutk3<-cutree(hc, k=3)
clustSize<-c(sum(cutk3==1),sum(cutk3==2),sum(cutk3==3))
print(clustSize)
[1] 400 400 200
[1] 196 204 400 200 # if cutree into 4 clusters

hc<-hclust(xx, method = "centroid", members=NULL)
plot(hc, labels=FALSE, xlab="", main="Dendrogram (centroid)", sub="")
cutk3<-cutree(hc, k=3)

clustSize<-c(sum(cutk3==1),sum(cutk3==2),sum(cutk3==3))
print(clustSize)
[1] 205 192 603
[1] 205 192 203 400 # if cutree into 4 clusters
```

6.3   K-means Cluster

6.3.1   General Description of K-means Clustering

K-means was first proposed by Steinhaus [82]. It is based on the least squares principle for multivariate data.

After pre-specifying the desired number of clusters (K), the principle is to find K "centroid point" locations in the data space, so that every observation is assigned to the closest centroid point in such a way that the sum of all Euclidean distances of the observations to the centroids is minimized. With this method, the centroid points are the means of observations assigned to them. The idea is that because all observations are as close as possible to their respective centroid, the clusters are very homogeneous; all cluster members can be well represented by their centroid. To be more specific, consider the following definition:

Definition Assume $U = \{x_1, x_2, \cdots, x_n\}$ being the data under consideration, let $d(x_i, x_j)$ be the distance between points $x_i, x_j \in U$. Then the K-means clustering of U is defined by first choosing $\{m_1, m_2, \cdots, m_K\}$ and $\{c_1, c_2, \cdots, c_K\}$ such that they minimize

$$S(m_1, m_2, \cdots, m_K) = \sum_{i=1}^{n} d(x_i, m_{c(i)}), \qquad (6.1)$$

where $c(i) = \min_{1 \leq j \leq K} d(x_i, m_j)$ and $\hat{m}_r = (1/n_r) \sum_{c(i)=r} x_i$.

Dendrogram (single)

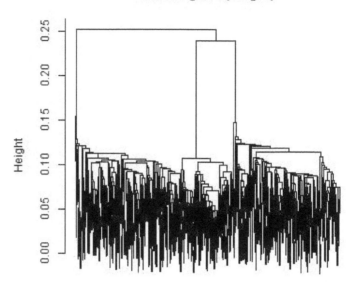

FIGURE 6.2
Hierarchical clustering of cassini data: single linkage.

K-means can also be derived as a maximum likelihood (ML) estimator in a particular statistical model. The K cluster centroid mean vectors are ML-estimators in a model in which the data are generated independently by K Gaussian distributions with different mean vectors. The covariance matrices need to be assumed to be equal and being a p-dimensional diagonal matrix multiplied with proper scale. Namely, K-means is implicitly based on Gaussian clusters with equal and spherical covariance matrices. In addition, K-means is not scale equivariant.

For data exploratory purpose, even if the above criteria are not sure to have been met (tests of equality of matrices do not usually have high power), one can still perform K-means clustering to the data, whether this corresponds to an intuitive clustering will need to be carefully examined so that refinement

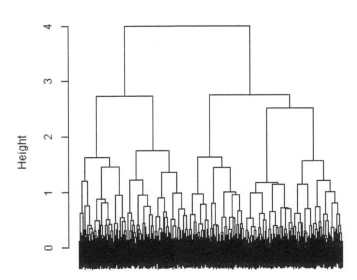

FIGURE 6.3
Hierarchical clustering of cassini data: complete linkage.

can be further perform. As a general practice of exploratory analysis, one often needs to repeat the data analysis in as many times as needed to arrive some satisfactory conclusion.

6.3.2 Estimating the Number of Clusters

To conduct the K-means clustering, analysts need to pre-specify the number of clusters, namely, the number K. Unfortunately, the number K is most likely unknown to the analysts until after a few trial and error runs. One may use the magnitude of equation (6.1) and pick the K which makes $S(m_1, m_2, \cdots, m_K)$ small. But the problem is that $S(m_1, m_2, \cdots, m_K)$ will decrease with the increasing number of clusters.

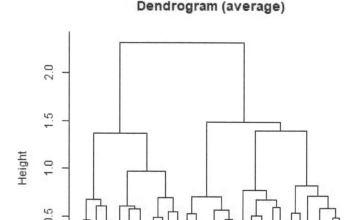

FIGURE 6.4

Hierarchical clustering of cassini data: average linkage.

One common but subjective approach is to use the *Scree Test*, which is a line plot with the value of $S(m_1, m_2, \cdots, m_K)$ against K and look for a clear break between the steep slope of the initial set of smaller K and the gentler slope of those K larger than these initial value of K. Unfortunately, interpretation of the plot is rarely as clear-cut as this, and in practice tends to involve a fairly subjective assessment of which values of K fall below an imaginary straight line extrapolated from the plots of the smaller K's. On the other hand, some analytical approaches had also been proposed in the literature, e.g., Sugar and James (2003); however, it may not provide straightforward solution. Ultimately, the most important point is to examine the clusters and consult with the subject experts to ensure the findings making sense.

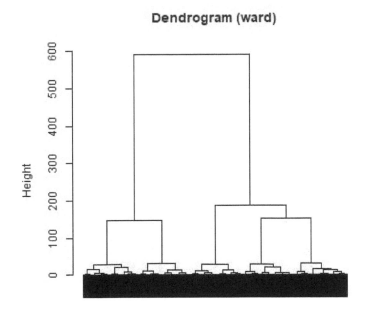

Dendrogram (ward)

FIGURE 6.5
Hierarchical clustering of cassini data: ward linkage

6.4 The PAM Clustering Algorithm

Similar to the K-means method, where the data are modeled using K cluster means that are iteratively refined by assigning all objects to the nearest mean and then update the mean of each cluster, Kaufman and Rousseeuw [42] proposed to use "K-medoids" approach and to "Partition Around Medoids" (PAM) with k representative objects m_i (called medoids) chosen from the data set to serve as "center" for the cluster with respect to the dissimilarity metrics. PAM consists of two algorithms BUILD and SWAP: using BUILD to choose an initial clustering and using SWAP to further improve the clustering toward a local optimum.

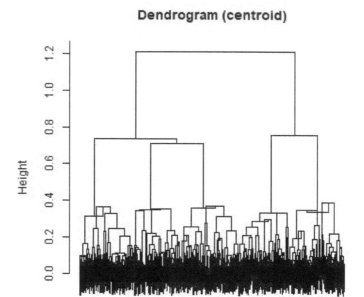

Dendrogram (centroid)

FIGURE 6.6
hierarchical clustering of cassini data: centroid linkage.

The PAM algorithm uses a greedy search which, however, may not find the optimum solution. It works as follows:

- Initialize: select k of the n data points as the medoids

- Associate each data point to the closest medoid.

- While the objective function (for minimizing the average dissimilarity) decreases: for each medoid m, for each non-medoid data point o:

  - Swap m and o, associate each data point to the closest medoid, recompute the cost (sum of distances of points to their medoid)
  - If the objective function of the configuration increased in the previous step, undo the swap

In general, PAM is more robust to noise and outliers as compared to k-means. Details can be found in [42]

6.4.1 Example of K-means with PAM Clustering Algorithm

```
pam.out =pam(RNA4clust, 3, diss=TRUE)
print("The medoids of the clusters")
print(pam.out$medoids)
[1]    1   2 102

print("The observation number of the medoids")
print(pam.out$id.med)
[1]    1   2 102

print("The cluster label of each observation")
print(pam.out$clustering) # this print the cluster membership for each case (omit).

print("The information of each cluster")
print(pam.out$clusinfo)
      size max_diss    av_diss diameter separation
[1,]   496 1.281181 -0.1640979 1.281181   -1.274127
[2,]   503 1.254352 -0.9545072 1.254352   -1.675547
[3,]     1 0.000000  0.0000000 0.000000   -1.675547
```

6.5 Bagged Clustering

Friedrich [46] proposed a cluster method based on bootstrap. The idea is to stabilize partitioning methods such as K-means by repeatedly running the cluster algorithm and combining the results. By repeatedly training on new data sets obtained from bootstrap, one gets different set of clusters which would be independent from random initializations. The centers of these clusters can be further segregated by hierarchical clustering, hence with their associated members of the original data.

Specifically, the bagged clustering algorithm works as follows:

- Construct B bootstrap training samples $\{X_1, \cdots, X_B\}$ by drawing with replacement from the original sample.

- Apply a cluster algorithm, e.g., K-means, to each set to obtain, a collection of $B \times K$ centers $C_B = \{c_{11}, \cdots, c_{1K}, c_{21}, \cdots, c_{BK}\}$, where K is the number of centers produced by the K-means cluster algorithm and c_{ij} is the j-th center using the ith bootstrap sample.

- Run a hierarchical cluster algorithm on C_B and let the center closest to x be $c(x) \in C_B$.

- A partition of the original data can be obtained by cutting the dendrogram at a certain level to obtain a partition $\{C_1, \cdots, C_m\}$, where $1 \leq m \leq BK\}$, of set C_B.

The author used the data `mlbench.cassini` from package `mlbench` in R to illustrate the application of the method. The data consist three well-defined clusters (see Figure 6.1) and the method was able to identify the clusters very well. The details are shown in the example section below.

6.5.1   Example of Bagged Clustering

Figure 6.1 is a simulated data `mlbench.cassini` with three well-defined clusters indicated by the different colors. Bagged clustering was tried with 3, 4, or 5 clusters and the sizes of the clusters indicated that 3 clusters are the most plausible classification and it reproduce the clusters (see Figure 6.7). The graph to help determine the number of clusters for Bagged clustering is shown in Figure 6.8.

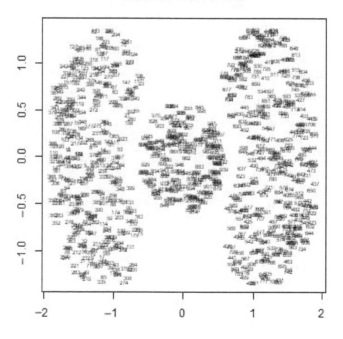

FIGURE 6.7
Bagged clustering of cassini data.

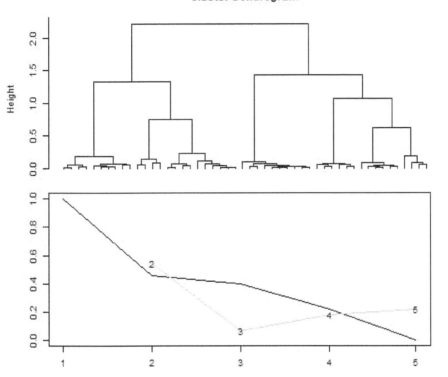

FIGURE 6.8
Cassini data cluster (bagged).

Other clustering methods were tested for the performance and the results are shown in Section 6.2.2. The R-program to illustrate the various cluster methods is shown below.

```
library(mlbench)
library(cluster)

# simulate cassini data
cassini<-mlbench.cassini(1000)
RNA4clust<-cassini$x
plot(RNA4clust[,1], RNA4clust[,2], main="cassini clusters", xlab="", ylab="",
col=cassini$classes)

# perform bagged clustering with 3 clusters
Kgrp<-3 or 4 or 5
Bclust <- bclust(RNA4clust, Kgrp, base.centers=5, docmdscale=TRUE, iter.base=10)
plot(Bclust)
plot(Bclust, labels=FALSE, xlab="", main="Dendrogram (bclust)", sub="")
```

```
Bcmd<-Bclust$cmdscale
plot(Bcmd[,1],Bcmd[,2], main="MDS of RNAseq data")

## outputs:
# bagged cluster of 3, 4, 5 clusters
   (1) size of 3 clusters: 347  56 113
   (2) size of 4 clusters:  75 311 129   1
   (3) size of 5 clusters:  83 320 108   3   2
```

6.6   RandomForest for Clustering

Even though `randomForest` has been commonly used for supervised data analysis; however, Brieman [37] also had discussed how to use it as an unsupervised tool to perform clustering. This approach is quite useful, especially if data have mixture of continuous and categorical values since it is not as easy to define distance between data points.

In unsupervised learning, the data consist of a set of x-vectors of the same dimension with no class labels or response variables. The approach in random forests is to consider the original data as class-1 and to create a synthetic second class of the same size that will be labeled as class-2. The synthetic class-2 is created by sampling at random from the univariate distributions of the original data. Specifically, the first coordinate is sampled from the N values $x(1, n)$. The second coordinate is sampled independently from the N values $x(2, n)$, and so forth.

Thus, class-2 has the distribution of independent random variables, each one having the same univariate distribution as the corresponding variable in the original data. Class-2 thus destroys the dependency structure in the original data. But now, there are two classes and this artificial two-class problem can be run through random forests. This allows all of the random forest options to be applied to the original unlabeled data set.

If the out-of-bag misclassification rate in the two-class problem is high, it implies that the x-variables look too much like independent variables to random forests. The dependencies do not have a large role and not much discrimination is taking place. On the other hand, if the misclassification rate is low, then the dependencies are playing an important role.

Formulating it as a two-class problem has a number of advantages. Missing values can be replaced effectively. Outliers can be found. Scaling can be performed (in this case, if the original data had labels, the unsupervised scaling often retains the structure of the original scaling), most importantly, it is the possibility of clustering.

6.6.1   Example: Random Forest for Clustering

```
library(mlbench)
library(cluster)
```

```
library(randomForest)

xx<-carssini4clust
xx<-xx[complete.cases(xx),]
xx<-as.matrix(xx)

fit1<-randomForest(xx, ntree=1000, keep.forest=FALSE, importance=TRUE, proximity=TRUE,
     na.action=na.omit)
print(fit1)
Call:  randomForest(x = xx, ntree = 1000, importance = TRUE, proximity = TRUE,
         keep.forest = FALSE, na.action = na.omit)
                   Type of random forest: unsupervised
                         Number of trees: 1000
No. of variables tried at each split: 1

# to examine clusters distribution
RFprox<-fit1$proximity
RFfit1.mds <- cmdscale(1 - fit1$proximity, eig=TRUE)
plot(RFfit1.mds$points[,1],RFfit1.mds$points[,2])
```

From the MDS plot in Figure 6.9, random forest produces a big clump of data points, which represent the middle cluster and the middle portion of the other two clusters. It is not very clear in cluster separation; however, it does give some indication of the proximity of the data points.

6.7 Mixture Models/Model-based Cluster Analysis

Mixture models (or model-based cluster analysis) are a clustering approach based on statistical models.

The general idea of mixture models is that the data are generated according to distributions from a parametric family $F = \{P_\theta$ with density $f_\theta, \theta \in \Theta\}$ and the data in different clusters follow distribution with different θ.

Specifically, data $\{x_1, x_2, \cdots, x_n\}$ are i.i.d. with density

$$f_\eta(x) = \sum_{k=1}^{K} \pi_k f_{\theta_k}(x)$$

where $0 \leq \pi_k \leq 1$ and $\sum_{k=1}^{K} \pi_k = 1$.

Let $\{\gamma_1, \gamma_2, \cdots, \gamma_n\}$ be the cluster labels for data $\{x_1, x_2, \cdots, x_n\}$, then

$$\gamma_i \sim \text{Multinormial}(1, \pi_1, \cdots, \pi_n)$$

and

$$x_i | (\gamma_i = k) \sim f_{\theta_{\gamma_i}}.$$

Therefore, the probability that x_i is classified in the kth class can be estimated as

$$p_{ik} = P(\gamma_i = k | x_i) = \frac{\pi_k f_{\theta_k}(x_i)}{\sum_{j=1}^{K} \pi_j f_{\theta_j}(x_i)},$$

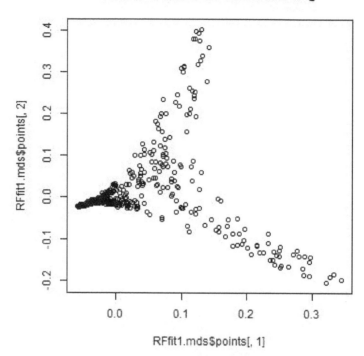

FIGURE 6.9
MDS plot from `randomForest` clustering of cassini data.

and the final assignment of cluster label for x_i can be determined by

$$\hat{\gamma}_i = \max_{k=1,\cdots,K} \hat{p}_{ik}.$$

If f_{θ_k} follows Gaussian distribution, then one would have the Gaussian mixture models. Since empirically almost every distribution can be approximated by a set of numerous Gaussian distributions, which can be considered as separate clusters, the number of clusters K can sometimes be over-estimated in this way. A more conservative approach is to use the Bayesian Information Criteria (BIC), which takes into account the magnitude of the likelihood function and also penalizes the excessive complexities of the models.

6.8 Stability of Clusters

As the outcomes of cluster analysis depend on several important factors, such as the measures of similarity, the pre-specification of the number of clusters such as in K-means clustering, the linkage methods, the choices of data standardization, and the nature of data features, etc., it is therefore important to validate the clusters by investigating the stability of the clusters besides the proper interpretations of the findings.

It is not unusual for clustering algorithms to produce several clusters with plausible interpretations of the underline rationales among the cluster members; however, it is also not unusual to have clusters with less commonality among the cluster members which will make good interpretation more difficult. These scenarios can be the results of clustering algorithms or the reflection of potential data issues and its stability.

Methods had been proposed in the literature, e.g., Hennig [33] and Fang and Wang [20] as well as the references therein to look into these issues. Hennig proposed bootstrap the Jaccard coefficient, sub-setting data, or introducing noise into data to investigate the cluster stability. Fang and Wang [20] also proposed to investigate cluster stability using bootstraps. To analyze the cluster stability, the `fpc` package in R has a function `clusterboot()` that uses bootstrap resampling for this purpose.

6.9 Consensus Clustering

In the general cluster analysis, it is not easy to determine the number of clusters and assign confidence to the selected number of clusters, as well as to the induced cluster assignments.

Monti et al. [57] and Wilkerson et al. [99] proposed a resampling-based method for class discovery. They use a subsampling techniques, whereby a subset of items is sampled without replacement from the original data set, to develop a general, model-independent resampling-based methodology of class discovery and clustering validation and visualization tailored to the task of analyzing big data, especially gene expression data. The authors call the methodology consensus clustering, which represent the consensus across multiple runs of a clustering algorithm, to determine the number of clusters in the data, and to assess the stability of the discovered clusters.

Specifically, for each pair of variables, the proportion of clustering runs in which two variables are clustered together is calculated. The consensus matrix, $\mathcal{M}$, is constructed by taking the average over the matrices of every bootstrapped sample.

Let D be the original data set, and $D_1, D_2, \cdots, D_h$ be the list of h sub-samples from D. Let M_h, an $N \times N$ matrix, corresponding to data set D_h. The entries of M_h are defined as follows:

$$M_h(i,j) = \begin{cases} 1 & \text{if items } i \text{ and } j \text{ belong to the same cluster} \\ 0 & \text{otherwise.} \end{cases} \quad (6.2)$$

Additionally, let matrix I_h, an $N \times N$ indicator matrix such that $I_h(i,j) = 1$ if both variables i and j are present in the data set D_h and 0 otherwise.

Then the consensus matrix $\mathcal{M}$ can then be defined based on $\{D(h) : h = 1, 2, ..., H\}$ as:

$$\mathcal{M}(i,j) = \sum_h M_h(i,j) / \sum_h I_h(i,j).$$

That is, the entry (i,j) in the consensus matrix records the number of times items i and j are assigned to the same cluster divided by the total number of times both items are selected. As defined, each entry in $\mathcal{M}$ is a real number between 0 and 1, and perfect consensus corresponds to a consensus matrix $\mathcal{M}$ with all the entries equal to either 0 or 1.

6.9.1   Determination of Clusters

The consensus matrix can be used as a visualization tool to help assess the clusters' composition, to assess the stability of the putative clusters, as well as their optimal number. The idea is to construct a consensus matrix $M(k)$ for each of a series of cluster numbers $\mathcal{K} = \{2, 3, \cdots, K\}$, to compare the resulting consensus matrices, and to select the cluster number based on the degree of scattering of the matrix.

For $0 \leq c \leq 1$, define the empirical cumulative distribution function as

$$CDF(c) = \sum_{i<j} I\{\mathcal{M}(i,j) \leq c\} / \{N(N-1)/2\}$$

The for each $k \in \mathcal{K}$, define the area under the CDF curve as

$$A(k) = \sum_{i=2}^{m} (x_i - x_{i-1}) CDF(x_i)$$

where the set $\{x_1, x_2, \cdots, x_m\}$ is the sorted set of entries of the consensus matrix $\mathcal{M}(k)$ and $m = N(N-1)/2$.

By plotting the relative increase of $A(k)$ for $k = 2, 3, \cdots, K$:

$$\Delta(k) = \begin{cases} A(k) & k = 2 \\ (A(k+1) - A(k))/A(k) & \text{otherwise.} \end{cases} \quad (6.3)$$

6.9.2 Example of Consensus Clustering on RNA Sequence Data

To demonstrate how consensus clustering can be performed, a RNA sequence from a recent study is used. The R-codes and the graphical outputs are shown below.

```
library(ConsensusClusterPlus)

d<-as.matrix(RNAseq.dat)
results = ConsensusClusterPlus(d,maxK=6,reps=50,pItem=0.8,pFeature=1,
        clusterAlg="hc",distance="pearson", seed=100,plot="jpeg")
```

The program produces clusters based on the parameter (maxK=·) specified, and in this case, it produces $2, 3, \cdots, 6$ clusters with clustering trees and heatmaps in Figures 6.10 to 6.12. By examining these figures, it is quite obvious that there are only two large clusters with a few "small and outlying" clusters. The program also produces cluster membership, which can be used to further examine the homogeneity of the individual clusters. As always, it is important to find a "physical meaning" of each cluster so that it can be easier for interpretations.

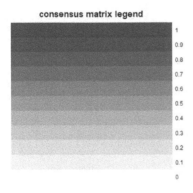

FIGURE 6.10
Legend and Heat Map for 2 classes.

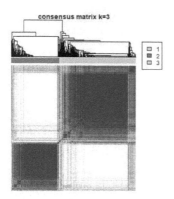

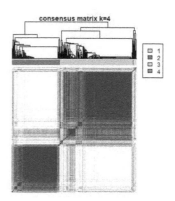

FIGURE 6.11
Heat Map for 3 and 4 classes.

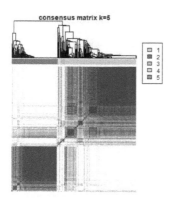

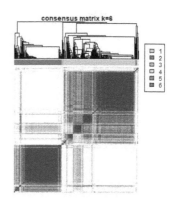

FIGURE 6.12
Heat Map for 5 and 6 classes.

To help determining the number of clusters, the program also produces the "CDF" and the area under the curve, which can be compared to examine whether there is any substantial difference between the areas under different curves to help decide the number of clusters. The delta curve plot as defined in Equation (6.3) is also provided to help determining the number of clusters by examine the drop of slop at adjacent points of the plot (Figure 6.13).

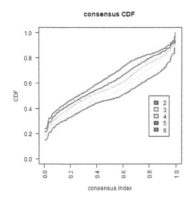 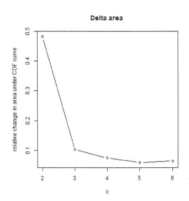

FIGURE 6.13
CDF of classes and difference of AUC.

6.10   The Integrative Clustering Framework

In genomic research, microarray-based platforms can generate various types of data, for example, continuous data for gene expression, DNA copy number, and CpG site methylation, while sequencing-based platforms can generate count data for gene expression and binary data for DNA mutation. Mo et al. [55] and [56] proposed a method to integrate data of various types, which represent the major forms of omics data. The same approach is applicable to any areas with various kinds of large data sets.

Let x_{ijt} denote the genomic variable associated with the jth ($j \in 1, \cdots, p_t$) genomic feature in the ith ($i \in 1, \cdots, n$) sample of the tth ($t \in 1, \cdots, m$) data type, and $z_i = (z_{i1}, \cdots, z_{ik})^t$ be a column vector consisting of k unobserved latent variables and $z_i \sim N(0, I_k)$.

The idea is to use a set of latent variables representing k distinct driving factors, which predict the values of the original $p = \sum_t p_t$ genomic variables, and collectively capture the major biological variations observed across cancer genomes. The genomic variables x_{ijt} are connected to the latent process via a parametric joint model in which different genomic variables are correlated through z_i.

As in the generalized linear model settings, which deals with data of various kinds, if x_{ijt} is a continuous variable, it is assumed that

$$x_{ijt} = \alpha_{jt} + \beta_{jt} + \epsilon_{ijt},$$

where $\epsilon_{ijt} \sim N(0, \sigma_{jt}^2)$.

If x_{ijt} is a binary variable, it is assumed that logit transformation can relate the probabilities with the linear model as in

$$\log \frac{P(x_{ijt} = 1 \mid z_i)}{1 - P(x_{ijt} = 1 \mid z_i)} = \alpha_{jt} + \beta_{jt}.$$

If x_{ijt} is a multicategory variable, it is assumed that the probability of each category is proportional to all categories after exponential transformation

$$P(x_{ijt} = 1 \mid z_i) = \frac{\exp(\alpha_{jct} + \beta_{jct} z_i)}{\sum_{l=1}^{C} \exp(\alpha_{ijt} + \beta_{ijt} z_i)},$$

where $P(x_{ijt} = \gamma \mid z_i)$ denote the probability of the γth category.

If x_{ijt} is a count variable, it is assumed that the log transformation will relate the count to the covariates

$$\log(\lambda(x_{ijt} \mid z_i)) = \alpha_{jt} + \beta_{jt} z_i$$

where $(\lambda(x_{ijt} \mid z_i)$ is the conditional mean of the count given z_i.

After the proper transformations of each data type and assume $\mathcal{L}(\cdots)$ being the likelihood function and also assuming the sparsity property of the model, the parameters can be estimated by maximizing the following equation

$$\max_{\alpha_{jt}, \beta_{jt}} \mathcal{L}(x_{ijt}, z_i; \alpha_{jt}, \beta_{jt}) - \sum_{t=1}^{m} \sum_{j=1}^{p_t} \lambda_t \|\beta_{jt}\|_1. \tag{6.4}$$

6.10.1   Example: Integrative Clustering

This section illustrates the procedures to perform integrative clustering using a brain tumor data (GBM) from TCGA database. Three kinds of data sets are used: GBM mutation (binary), Copy Numbers (count), and gene expression data (continuous). The R-codes are shown below:

```
library(iClusterPlus)
data(gbm)

# input parameters:
gbm.mut    # mutation data
gbm.cn     # copy number
```

```
gbm.exp    # gene expression dat
numCov     # number of covariates
numCluster=numCov+1
lambdaValue=c(0.02, 0.05, 0.08)  ## set the various lambda values for the model

fit.single=iClusterPlus(dt1=gbm.mut2,dt2=gbm.cn,dt3=gbm.exp, type=c("binomial","possion","gaussian"),
                lambda=lambdaValue, K=numCov, maxiter=10)

# the output variables
print(attributes(fit.single))
$names
[1] "alpha"      "beta"       "clusters" "centers"    "meanZ"      "BIC"        "dev.ratio" "dif"

# alpha has 3 block ([[1]], [[2]], [[3]]) corresponding to the 3 datasets with length dt1, dt2, dt3
print(fit.single$alpha[[1]])

 [1] -3.295837  0.000000 -3.470615 -3.713572 -3.474837 -2.995732 -2.805875 -2.643916  0.000000 ...

# beta has 3 two-dimensional block with nrow dt1, dt2, dt3
print(fit.single$beta[[1]])
              [,1]           [,2]
 [1,]   0.0000000  0.0000000000
 [2,]   0.0000000  0.0000000000
 [3,]   0.0000000 -0.4364227834
 [4,]   0.0000000  0.0000000000
 [5,]   0.0000000 -0.4409437006
 [6,]   0.0000000  0.0000000000
 [7,]   0.0000000 -0.2662370949
 [8,]   0.0000000 -0.3478079025
 [9,]   0.0000000  0.0000000000
[10,]   0.0000000  0.0000000000
 .....
# note: due to sparsity, many features have value zero.

# print the cluster assignment for each sample/subject
print(fit.single$clusters)  # since K=2, it produces 3 clusters:

 [1] 3 1 1 1 3 1 1 3 1 3 1 1 3 1 1 3 1 1 1 3 1 1 1 3 3 3 3 1 3 ...

print(fit.single$meanZ)  # the estimated values for the latent variables.
            [,1]         [,2]
 [1,]   0.31750111  0.67572158
 [2,]  -0.66688062 -0.45035373
 [3,]  -1.33399671  2.80217699
 [4,]  -1.44404080  0.75448553
 [5,]   0.46772169  0.04515254
 [6,]  -0.74479558 -1.78891194
 [7,]  -0.19915774  0.84569384
 [8,]   0.36563617 -1.49382822
 [9,]  -0.22389096  0.59032681
[10,]   0.88109524 -0.05339424
 .....
```

After the analysis, one can perform further analysis, e.g., fine tune the model or selection of features, for each cluster. One can also examine the "tightness" of each cluster to better understand the homogeneity of the cluster members for each data set.

7

Neural Network

Neural networks are inspired by the way the human brain works. A human brain can process huge amounts of information using data sent by human senses. The processing is performed by neurons, which work on electrical signals passing through them and applying flip-flop logic, like opening and closing of the gates for signal to transmit through.

7.1 General Theory of Neural Network

Neural networks are mathematical models represented by a collection of simple computational units interlinked by a system of connections. The number of units can be very large and the connections intricate. Over the years, neural networks had been used for many applications such as pattern recognition, etc. Figure 7.1 shows the so-called single unit perceptron, which was first proposed by a neuro-scientist Rosenblatt (1965) with the purposes to construct a system that could compute, learn, remember, and optimize in the same way as a human brain.

A perceptron is a single neuron that classifies a set of inputs into one of two categories (usually 1 or -1). If the inputs are in the form of a grid, a perceptron can be used to recognize visual images of shapes. The perceptron usually uses a step function, which returns 1 if the weighted sum of the inputs exceeds a threshold and 0 otherwise. When layers of perceptron are combined together, they form a multilayer architecture, and this gives the required complexity of the neural network processing. Multi-Layer Perceptrons (MLPs) are the most widely used architecture for neural networks.

Mathematically, the percepton expresses the relationship between the input variables $\{x_1, x_2, \cdots, x_n\}$ and the output y:

$$y = w_0 + \sum_{j=1}^{n} w_j x_j = \sum_{j=0}^{n} w_j x_j$$

if we define $x_0 = 1$. In practices, the most common neural-network approach is multilayer perceptrons and generalizations of single-layer perceptrons.

DOI: 10.1201/9781003205685-7

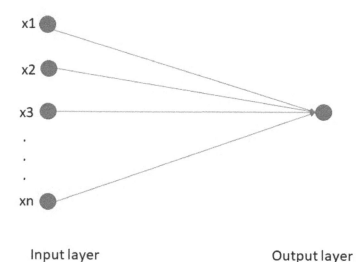

Input layer   Output layer

FIGURE 7.1
Structure of Perceptron.

7.2 Elemental Aspects and Structure of Artificial Neural Networks

McCulloch and Pitts (1943) published a seminal paper on artificial intelligence sought to discover why and how the human brain processes or cognitive tasks.

For a set of input data $\{x_1, x_2, \cdots, x_n\}$, they proposed a binary function

$$Y = f\left(\sum_{j=1}^{n} w_j x_j + w_0\right) \tag{7.1}$$

with function f being a binary indicator of positive or negative of the quantity, the $\{w_j, j = 1, \cdots, n\}$ being connection weights, connection strengths, or connectivities, and w_0 being the bias term using neurological terminology.

One can rewrite equation (7.1) in a more general form as

$$y = f(\phi(x, w)), \tag{7.2}$$

where f and ϕ are prescribed functional forms, x and w represent the inputs data and the connection weights, respectively. The function f is called the activation function and usually, ϕ is linear as in equation (7.1). The following is a list of commonly used activation functions:

- $f(u) = u$.

- $f(u) = \mathrm{sgn}(u) = f_h(u)$, the hard limiter nonlinearity, produces binary $(+1/-1)$ output.

- $f(u) = (1 + e^{-u})^{-1} = f_s(u)$ the sigmoidal (logistic) nonlinearity, produces output between 0 and 1.

- $f(u) = \tanh(u)$ produces output between -1 and 1.

- $f(u) = (u)_+$ produces a non-negative output.

- $f(u) = +1$ with probability $f_s(u)$ and $f(u) = -1$ with probability $1 - f_s(u)$ provides random binary $(+1)$ output via logistic regression.

7.3 Multilayer Perceptrons

Networks are used in practice to process a set of items, which are associated with a vector of measurable features x and a target or response vector z, which represents, for instance, the indicator of the true speech pattern, digit, or disease category or a more general response. When the network receives the input vector x and creates a (set of) outputs y as a predictor of the unknown z. The output y y is a function of the network architecture, i.e., the set of activation functions and all the parameters.

7.3.1 The Simple (Single Unit) Perceptron

For the single-unit perceptron, the output y can be expressed as

$$y = f_h\left(\sum_{j=1}^{n} w_j x_j + w_0\right)$$

$$= f_h\left(\sum_{j=0}^{n} w_j x_j\right) \tag{7.3}$$

$$= f_h(w^t x).$$

7.3.2 Training Perceptron Learning

Training of the perceptron learning rule is usually via a recursive algorithm in which the weights are modified as the training data are processed. Given an observation (x, z) from the training set and that $y = y(w)$ denotes the (binary)

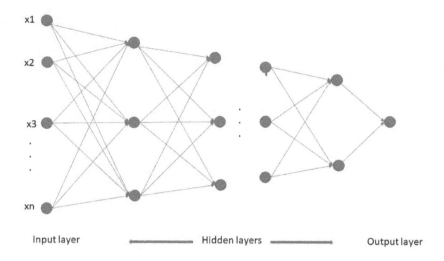

FIGURE 7.2
General structure of neural/deep learning network.

prediction for z, given x, based on the current values w for the weights and bias, and let $E(w)$ denote the prediction error. In order to minimize $E(w)$, one can use the gradient decent by taking derivative with respect to w_j for $j = 0, \cdots, n$, namely,

$$\Delta w_j = -\eta \frac{\partial E(w)}{\partial w_j}$$

and w_j can be updated according to

$$w_j \to w_j + \Delta w_j$$

and η, the so-called learning rate.

7.4   Multilayer Perceptrons (MLP)

7.4.1   Architectures of MLP

In general practice, simple perceptron can be quite limited to deal with more complicated problems. Multilayer perceptrons are extension of simple perceptron and are far more flexible for predictions. It usually consists of an input layer, one or more hidden layers, and one output layer (Figure 7.2).

The hidden layers are arranged in an one-directional sequence with the outputs from a previous layer serve as the inputs for the next layer. Consequently, the multilayer perceptron has a feed-forward network structure. Weights are specified for all connections. Biases and activation functions are proposed for each of the hidden and output nodes. The outputs need not be binary.

The mathematical description of MLP is actually quite straightforward by extending equation (7.2) of the structure of simple perceptron. Let

$$\Phi = (\phi_1, \phi_2, \cdots, \phi_K)$$

be the activation functions for the first to the Kth hidden layer, respectively, and the respective weight matrix by $\mathcal{W} = \{W_1, W_2, \cdots, W_K\}$ and input data matrix be X, then the outputs of the output layer can be expressed as

$$Y = \Phi(X, W_1, W_2, \cdots, W_K) = \phi_K(\cdots, \phi_2(\phi_1(X, W_1), W_2), \cdots, W_K) \quad (7.4)$$

7.4.2 Training MLP

Training of the MLP is similar to that for simple percepton with higher degree of complexities.

Given an observation (X, Z) from the training set and that Y as in equation (7.4) denotes the prediction for Z, given X, based on the current values $\mathcal{W}$ for the weights and bias, and let $E(\mathcal{W})$ denote the prediction error. In order to minimize $E(\mathcal{W})$, one can use the gradient descent by taking derivatives with respect to each element of $\mathcal{W}$, namely,

$$\Delta w_{ij} = -\eta_{ij} \frac{\partial E(\mathcal{W})}{\partial w_{ij}}$$

and w_{ij} can be updated according to

$$w_{ij} \rightarrow w_{ij} + \Delta w_{ij} \quad (7.5)$$

and η_{ij} being the learning rate.

In practice, numerical methods are required to minimize $E(\mathcal{W})$, and techniques such as conjugate gradients, quasi-Newton algorithms, simulated annealing, and genetic algorithms have been used. These methods are often much faster than the method of error back-propagation.

7.5 Deep Learning

Deep network learning (DL) forms an advanced neural network with numerous hidden layers (Figure 7.2). DL is a vast subject and is an important concept for building Artificial Intelligence (AI). It is used in various applications, such as: Image recognition, computer vision, Handwriting detection, Text classification, Multiclass classification, Regression problems, and more.

With the advent of big data processing infrastructure, GPU, and GP-GPU, it can overcome the challenges with shallow neural networks, namely, overfitting and vanishing gradient, using various activation functions and L1/L2 regularization techniques. Deep learning can work on large amounts of labeled and unlabeled data easily and efficiently.

Deep learning is a class of machine learning, wherein learning happens on multiple levels of neuron networks. The standard diagram depicting a DNN is shown in the following figure:

From the analysis of the previous figure, we can notice a remarkable analogy with the neural networks we have studied so far. We can then be quiet, unlike what it might look like, deep learning is simply an extension of the neural network. In this regard, most of what we have seen in the previous chapters is valid. In short, a DNN is a multilayer neural network that contains two or more hidden layers. Nothing very complicated here. By adding more layers and more neurons per layer, we increase the specialization of the model to train data but decrease the performance on the test data.

7.5.1   Model Parameterization

Like in many statistical modeling, often there are more than one model to fit the data to increase the goodness of fit. Similarly, in neural network modeling, one can construct better models by specifying the structure of network with proper specification of the model parameters. It takes a lot of experience and intuition to find the a suitable network for good predictions. This can depend on the optimal values of the parameters of the model, which must be specified before starting the training process so that the models train better and more quickly. In the following, we describe some parameters, which one can specify for a better network model.

- Initialization of parameter weights: The initialization of the parameters' weight is important for the efficiency of network training. It is advisable to initialize the weights with random values to break the symmetry between different neurons. It is also important to avoid the numbers which are either too small or too large.

- Activation function: In general, heuristics can be used to select the type of activation functions. It is advisable to try different functions and taking into account the data types as well as the objectives of the analyses.

- Determine the number of neurons or layers: One can add more neurons in a layer or add more layers; however, in these cases, the gains in accuracy have the cost of increasing the execution time of the learning process.

- Increase epochs: Epochs determine the number of times all the training data have passed through the neural network in the training process. One can increase the number of epochs, however, beware of the overfitting which will affect the prediction performance for new data.

- Bath size: One can partition the training data in mini batches to pass them through the network. For example, in R-package `keras`, the `batch_size` is the argument that indicates the size of these batches that will be used in the `fit()` method in an iteration of the training to update the gradient.

- Learning rate, the magnitude of the gradient descent multiply the learning rate will determine the step size of the next point in backpropagation (see equation (7.5)). The proper value of this parameter is very much dependent on the problem in question, but in general, if this is too big, huge steps are being made, which could be good to go faster in the learning process. One may not want to hold the learning rate constant. It is advisable to decrease the learning rate when the model approaches a solution with minimum error.

- Momentum: In the processes to minimize the error function, instead of finding the solution for global minimum, the optimizer may find a solution for local minimum only. One way to solve this situation could be to restart the process from different random positions and, in this way, increase the probability of reaching the global minimum. An alternative is to specify the momentum parameter by weighting the previous gradients to obtain a constant between 0 and 1 to be assigned as the momentum.

In R-package `keras`, which provides specifications for describing dense neural networks, convolution neural networks and recurrent neural networks running on top of either "TensorFlow" or "Theano," one needs to specify these parameters. For example, in the function for stochastic gradient descent optimization, the syntex of the function `optimizer_sgd(lr =, momentum =, decay =, nesterov =, clipnorm =, clipvalue =)`. Analysts are advised to try various values for these parameters. R also has package `kerasR`, which provides a consistent interface to the keras Deep Learning Library directly from within R.

7.6   Few Pros and Cons of Neural Networks

Neural networks are flexible and can be used for both regression and classification problems. Any data that can be made numeric can be used in the model, as neural network is a mathematical model with approximation functions. Neural networks are good to model with nonlinear data with a large number of inputs; for example, images. It is reliable in an approach of tasks involving many features. It works by splitting the problem of classification into a layered network of simpler elements. Neural networks can be trained with any number of inputs and layers. Neural networks work best with more data points. Once trained, the predictions are pretty fast.

On the other hand, neural networks are "black boxes," it is not easy for one to know how much each independent variable is influencing the dependent variables. This is quite different from the usual regression modeling, which one can test the significance of each covariate.

Neural networks are computationally very expensive and time consuming to train with traditional CPUs. Like any statistical analyses, the performance and capability of generalization of neural networks depend on training data. In general, one needs to be aware of the possibility of over-fitting. As a general rule of thumb, to ensure the good performance of generalizability, large sample of representative data is necessary to train the network.

7.7 Examples

In this section, the Boston Housing data have been used to illustrate the application of single-layer, two-layer network. The R-package used is **neuralnet**. For the two-layer network, two examples are shown with different number of internal nodes in the hidden layer. The performances of these variations of network structures are somewhat different. As mentioned above, the network structure can depend on the nature of the data and specification of parameters. The more complicated network structure does not necessary imply better prediction performance. The outputs and figures for the following example of networks are shown below.

```
library(neuralnet)
library(MASS)
set.seed(1)
data = Boston
max_data <- apply(data, 2, max)
min_data <- apply(data, 2, min)
data_scaled <- scale(data,center = min_data, scale = max_data - min_data)
index = sample(1:nrow(data),round(0.70*nrow(data)))
train_data <- as.data.frame(data_scaled[index,])
test_data <- as.data.frame(data_scaled[-index,])
n = names(data)
f = as.formula(paste("medv ~", paste(n[!n %in% "medv"], collapse = " + ")))
#net_data = neuralnet(f,data=train_data,hidden=c(10),linear.output=T)
#net_data = neuralnet(f,data=train_data,hidden=c(5,5),linear.output=T)
net_data = neuralnet(f,data=train_data,hidden=c(6,4),linear.output=T)
print(net_data)
plot(net_data)

predict_net_test <- compute(net_data,test_data[,1:13])
predict_net_test_start <- predict_net_test$net.result*(max(data$medv)
                             -min(data$medv))+min(data$medv)
test_start <- as.data.frame((test_data$medv)*(max(data$medv)
```

```
                                        +min(data$medv))
MSE.net_data <- sum((test_start - predict_net_test_start)^2)/nrow(test_start)
Regression_Model <- lm(medv~., data=data)
print(summary(Regression_Model))
test <- data[-index,]
predict_lm <- predict(Regression_Model,test)
MSE.lm <- sum((predict_lm - test$medv)^2)/nrow(test)
print(MSE.net_data)
print(MSE.lm)
```

```
$result.matrix
                                [,1]
error                    2.673335e-01
reached.threshold        9.417018e-03
steps                    1.717800e+04
Intercept.to.1layhid1   -5.885270e-01
crim.to.1layhid1         2.610908e+00
zn.to.1layhid1          -3.382429e-01
indus.to.1layhid1       -6.723215e-01
chas.to.1layhid1        -2.141552e+00
nox.to.1layhid1          6.505496e-01
rm.to.1layhid1          -8.412213e-01
age.to.1layhid1         -1.121926e-01
dis.to.1layhid1          2.357168e+00
rad.to.1layhid1          1.856039e-01
tax.to.1layhid1         -7.461289e-01
ptratio.to.1layhid1     -9.661230e-01
black.to.1layhid1        1.086747e+00
lstat.to.1layhid1        5.797324e+00
Intercept.to.1layhid2    9.987863e-01
crim.to.1layhid2         5.574059e+00
zn.to.1layhid2           3.479829e+00
indus.to.1layhid2       -2.680882e-01
chas.to.1layhid2        -8.935932e-02
nox.to.1layhid2         -5.854433e+00
rm.to.1layhid2           6.958981e-01
age.to.1layhid2          3.843987e-01
dis.to.1layhid2         -5.722064e+00
rad.to.1layhid2         -3.930539e-01
tax.to.1layhid2          1.126316e+00
ptratio.to.1layhid2     -1.006863e+00
black.to.1layhid2       -8.866398e-01
lstat.to.1layhid2        8.604476e-01
Intercept.to.1layhid3   -4.388686e-01
crim.to.1layhid3        -4.514075e+00
zn.to.1layhid3          -6.644275e-01
indus.to.1layhid3       -7.698197e-01
chas.to.1layhid3         8.956544e-01
nox.to.1layhid3          2.566014e+00
rm.to.1layhid3           1.230372e+00
age.to.1layhid3         -4.528008e-01
dis.to.1layhid3          2.101501e+00
rad.to.1layhid3          5.641753e+00
tax.to.1layhid3         -4.615360e-01
ptratio.to.1layhid3     -3.044266e-01
```

```
black.to.1layhid3          5.624067e-01
lstat.to.1layhid3          1.936569e-01
Intercept.to.1layhid4     -2.510902e+00
crim.to.1layhid4           4.199632e+00
zn.to.1layhid4            -2.075045e-01
indus.to.1layhid4         -7.930703e-01
chas.to.1layhid4          -2.251177e+02
nox.to.1layhid4           -5.795552e-01
rm.to.1layhid4             1.703063e-01
age.to.1layhid4            7.403411e-01
dis.to.1layhid4           -7.896782e-01
rad.to.1layhid4            1.529801e+00
tax.to.1layhid4            3.063752e-01
ptratio.to.1layhid4       -9.342721e-01
black.to.1layhid4          6.987342e-01
lstat.to.1layhid4          1.764579e-01
Intercept.to.1layhid5      1.528322e+00
crim.to.1layhid5          -1.858837e+02
zn.to.1layhid5            -1.849334e+00
indus.to.1layhid5         -8.018930e-01
chas.to.1layhid5           1.386731e+00
nox.to.1layhid5            1.030169e+00
rm.to.1layhid5             2.661699e+00
age.to.1layhid5           -3.001826e-01
dis.to.1layhid5           -2.850614e+00
rad.to.1layhid5            1.399619e+00
tax.to.1layhid5           -4.988481e+00
ptratio.to.1layhid5       -1.908737e+00
black.to.1layhid5         -1.298662e+00
lstat.to.1layhid5          3.769287e-01
Intercept.to.1layhid6     -2.275486e+00
crim.to.1layhid6           1.356385e+00
zn.to.1layhid6            -8.983661e-02
indus.to.1layhid6         -5.051195e-01
chas.to.1layhid6           2.287640e+00
nox.to.1layhid6            6.935143e-01
rm.to.1layhid6             5.020964e-01
age.to.1layhid6           -2.524577e-01
dis.to.1layhid6           -1.318093e+00
rad.to.1layhid6            2.430276e-01
tax.to.1layhid6           -2.748716e-01
ptratio.to.1layhid6        9.104794e-01
black.to.1layhid6         -1.900373e+00
lstat.to.1layhid6         -6.692689e-01
Intercept.to.1layhid7     -3.913386e+00
crim.to.1layhid7           3.214788e+00
zn.to.1layhid7             2.257941e+00
indus.to.1layhid7         -8.095075e+00
chas.to.1layhid7           3.349163e+01
nox.to.1layhid7            3.936930e+01
rm.to.1layhid7            -4.216308e+00
age.to.1layhid7           -3.986298e+00
dis.to.1layhid7           -3.643969e+00
rad.to.1layhid7           -2.398723e+00
tax.to.1layhid7           -1.648353e+01
ptratio.to.1layhid7        7.149531e+00
black.to.1layhid7          1.090157e+01
```

```
lstat.to.1layhid7         1.780226e+01
Intercept.to.1layhid8     2.135229e-01
crim.to.1layhid8          1.684184e+02
zn.to.1layhid8           -2.575938e+00
indus.to.1layhid8        -7.834845e-01
chas.to.1layhid8         -2.788822e+00
nox.to.1layhid8           1.190700e+00
rm.to.1layhid8            1.257320e+00
age.to.1layhid8          -1.215297e+00
dis.to.1layhid8           2.939946e+00
rad.to.1layhid8          -5.402391e+00
tax.to.1layhid8          -7.896869e-02
ptratio.to.1layhid8       1.291753e-01
black.to.1layhid8         1.285225e+00
lstat.to.1layhid8         4.667606e+00
Intercept.to.1layhid9    -4.790393e+00
crim.to.1layhid9          1.204484e+02
zn.to.1layhid9            3.438245e+00
indus.to.1layhid9         3.473758e+00
chas.to.1layhid9          3.240575e+00
nox.to.1layhid9          -2.342448e+00
rm.to.1layhid9            3.195149e+00
age.to.1layhid9          -2.021393e+00
dis.to.1layhid9          -2.282079e-01
rad.to.1layhid9          -1.687336e+00
tax.to.1layhid9           3.057698e+00
ptratio.to.1layhid9       8.479316e-01
black.to.1layhid9        -5.102346e-01
lstat.to.1layhid9         2.369450e+01
Intercept.to.1layhid10   -1.159577e+00
crim.to.1layhid10        -1.606417e+01
zn.to.1layhid10           2.197682e+00
indus.to.1layhid10        3.166498e+00
chas.to.1layhid10        -9.832943e-01
nox.to.1layhid10          1.758357e+00
rm.to.1layhid10          -2.645068e+00
age.to.1layhid10         -3.642697e+00
dis.to.1layhid10         -2.434195e+00
rad.to.1layhid10         -2.732939e+00
tax.to.1layhid10          3.140089e+00
ptratio.to.1layhid10      1.067100e+00
black.to.1layhid10        5.310493e-01
lstat.to.1layhid10        2.551671e+01
Intercept.to.medv        -1.494619e+00
1layhid1.to.medv         -7.270609e-01
1layhid2.to.medv          6.635326e-01
1layhid3.to.medv          1.113972e+00
1layhid4.to.medv         -6.766912e-01
1layhid5.to.medv          7.294035e-01
1layhid6.to.medv         -2.326968e+00
1layhid7.to.medv          3.619904e-01
1layhid8.to.medv          1.114781e+00
1layhid9.to.medv          5.172887e-01
1layhid10.to.medv        -4.901125e-01

MSE.net_data
[1] 19.2383
```

```
MSE.lm
[1] 26.54428
```

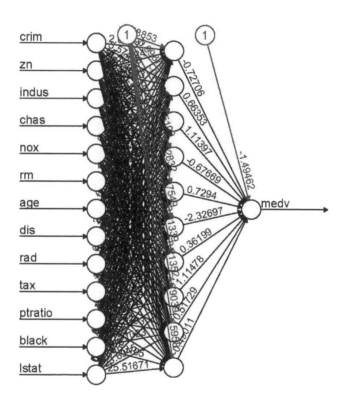

FIGURE 7.3
Neural network with 1 hidden layer of 10 nodes.

```
$result.matrix
                              [,1]
error                    3.271702e-01
reached.threshold        9.196184e-03
steps                    2.551000e+03
Intercept.to.1layhid1   -5.842410e-01
```

```
crim.to.1layhid1          9.482904e-01
zn.to.1layhid1           -8.537766e+00
indus.to.1layhid1         8.362651e-01
chas.to.1layhid1          8.640874e-01
nox.to.1layhid1          -1.197203e+01
rm.to.1layhid1            2.574914e+00
age.to.1layhid1           2.491167e+00
dis.to.1layhid1          -3.839449e+00
rad.to.1layhid1           1.958128e+00
tax.to.1layhid1          -1.691548e+00
ptratio.to.1layhid1      -2.446320e+00
black.to.1layhid1         5.038176e-01
lstat.to.1layhid1         2.596390e+00
Intercept.to.1layhid2     1.037490e+00
crim.to.1layhid2         -7.818992e+01
zn.to.1layhid2           -6.726457e-01
indus.to.1layhid2         3.963050e-01
chas.to.1layhid2          1.391128e-01
nox.to.1layhid2           2.196174e-02
rm.to.1layhid2           -2.614369e+00
age.to.1layhid2          -2.818530e-01
dis.to.1layhid2           2.221439e+00
rad.to.1layhid2           6.119317e+00
tax.to.1layhid2           2.129431e+00
ptratio.to.1layhid2      -5.753316e-01
black.to.1layhid2        -1.404752e+00
lstat.to.1layhid2         8.480773e+00
Intercept.to.1layhid3    -4.728816e-01
crim.to.1layhid3         -1.883340e+01
zn.to.1layhid3            1.623615e-01
indus.to.1layhid3         8.071993e-01
chas.to.1layhid3          2.182723e+00
nox.to.1layhid3           1.922534e+00
rm.to.1layhid3            2.627601e+00
age.to.1layhid3          -1.140245e+00
dis.to.1layhid3          -1.011386e+00
rad.to.1layhid3           7.419382e-01
tax.to.1layhid3          -8.623932e-01
ptratio.to.1layhid3      -4.675322e-01
black.to.1layhid3        -9.176692e-02
lstat.to.1layhid3        -1.592584e+00
Intercept.to.1layhid4    -7.318581e-01
crim.to.1layhid4          1.587127e+01
zn.to.1layhid4            8.582288e-01
indus.to.1layhid4         3.815123e+00
chas.to.1layhid4         -2.601803e+00
nox.to.1layhid4           1.255850e+00
rm.to.1layhid4           -4.882366e+00
age.to.1layhid4           9.712710e-01
dis.to.1layhid4           2.450454e+00
rad.to.1layhid4           4.058123e+00
tax.to.1layhid4          -1.654850e+00
ptratio.to.1layhid4      -1.286999e+00
black.to.1layhid4         1.358178e+00
lstat.to.1layhid4        -4.838777e+01
Intercept.to.1layhid5     2.435650e+00
crim.to.1layhid5         -3.132796e+00
```

```
zn.to.1layhid5            -4.085856e-02
indus.to.1layhid5          8.850446e-01
chas.to.1layhid5          -1.365535e+02
nox.to.1layhid5            4.717066e-01
rm.to.1layhid5            -3.346629e+00
age.to.1layhid5           -1.056865e+00
dis.to.1layhid5           -1.227830e+00
rad.to.1layhid5           -1.107752e+00
tax.to.1layhid5           -1.913230e+00
ptratio.to.1layhid5       -7.235855e-01
black.to.1layhid5          1.091631e+00
lstat.to.1layhid5         -7.144897e+00
Intercept.to.1layhid6     -2.196119e+00
crim.to.1layhid6          -5.251479e+00
zn.to.1layhid6            -1.036288e+00
indus.to.1layhid6         -2.497746e-01
chas.to.1layhid6           2.346139e+00
nox.to.1layhid6            2.876085e+00
rm.to.1layhid6             1.450717e-01
age.to.1layhid6            1.877575e+00
dis.to.1layhid6           -3.603918e-01
rad.to.1layhid6           -5.226998e+00
tax.to.1layhid6            1.558472e+00
ptratio.to.1layhid6        1.272218e+00
black.to.1layhid6         -7.222035e-01
lstat.to.1layhid6         -1.214468e+00
Intercept.to.2layhid1     -3.434959e-01
1layhid1.to.2layhid1      -1.713829e+00
1layhid2.to.2layhid1      -2.230749e+00
1layhid3.to.2layhid1      -1.287205e+00
1layhid4.to.2layhid1       1.928565e+00
1layhid5.to.2layhid1       3.013110e+00
1layhid6.to.2layhid1       2.276591e-01
Intercept.to.2layhid2     -2.213322e-01
1layhid1.to.2layhid2       4.809544e+00
1layhid2.to.2layhid2       1.629501e+00
1layhid3.to.2layhid2       3.330490e-01
1layhid4.to.2layhid2      -2.013759e+00
1layhid5.to.2layhid2       9.368345e+01
1layhid6.to.2layhid2      -2.283039e-01
Intercept.to.2layhid3     -1.435934e-01
1layhid1.to.2layhid3      -1.472804e-01
1layhid2.to.2layhid3      -2.072495e+00
1layhid3.to.2layhid3       1.366656e+00
1layhid4.to.2layhid3       2.688446e+00
1layhid5.to.2layhid3      -2.170495e-01
1layhid6.to.2layhid3      -9.750202e-01
Intercept.to.2layhid4     -4.253506e-01
1layhid1.to.2layhid4       9.536135e-01
1layhid2.to.2layhid4       7.187264e-01
1layhid3.to.2layhid4       5.193062e-01
1layhid4.to.2layhid4       8.899697e-01
1layhid5.to.2layhid4       1.716246e-02
1layhid6.to.2layhid4      -2.283689e-01
Intercept.to.medv         -7.676458e-01
2layhid1.to.medv          -1.395794e+00
2layhid2.to.medv           9.998766e-01
```

```
2layhid3.to.medv        2.066817e+00
2layhid4.to.medv       -1.272764e-01

MSE.net_data
[1] 18.08116

MSE.lm
[1] 26.54428
```

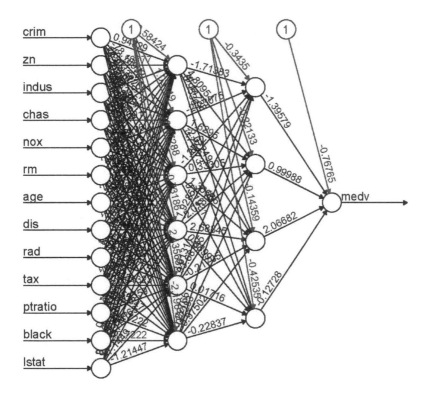

FIGURE 7.4
Neural network with 2 hidden layers of 6 and 4 nodes.

```
$result.matrix
                   [,1]
```

```
error                       4.634053e-01
reached.threshold           9.931643e-03
steps                       2.610000e+03
Intercept.to.1layhid1      -6.299284e-01
crim.to.1layhid1            1.244527e+00
zn.to.1layhid1             -6.204187e-01
indus.to.1layhid1          -9.784589e-01
chas.to.1layhid1           -1.037439e-01
nox.to.1layhid1             2.883637e-01
rm.to.1layhid1              1.595316e+00
age.to.1layhid1            -4.689363e-02
dis.to.1layhid1            -1.722548e+00
rad.to.1layhid1             2.665349e+00
tax.to.1layhid1             5.503176e-01
ptratio.to.1layhid1        -4.323036e-01
black.to.1layhid1           1.116121e+00
lstat.to.1layhid1          -2.282447e-01
Intercept.to.1layhid2       6.620853e-01
crim.to.1layhid2           -5.008554e+01
zn.to.1layhid2              9.616267e-01
indus.to.1layhid2          -1.614981e+00
chas.to.1layhid2            8.189316e-01
nox.to.1layhid2             4.501484e+00
rm.to.1layhid2              1.339490e+00
age.to.1layhid2            -1.692416e+00
dis.to.1layhid2            -5.429183e-01
rad.to.1layhid2             3.862691e+00
tax.to.1layhid2             6.545936e-01
ptratio.to.1layhid2        -3.080457e+00
black.to.1layhid2          -1.545006e+00
lstat.to.1layhid2          -1.405028e+01
Intercept.to.1layhid3      -9.445828e-01
crim.to.1layhid3           -6.512463e+00
zn.to.1layhid3              8.109315e-01
indus.to.1layhid3           1.123383e-01
chas.to.1layhid3           -9.100855e-03
nox.to.1layhid3             1.450572e+00
rm.to.1layhid3              1.054945e+00
age.to.1layhid3            -7.502149e-01
dis.to.1layhid3            -5.545173e+00
rad.to.1layhid3             2.393718e+00
tax.to.1layhid3             1.021356e-01
ptratio.to.1layhid3        -1.531381e+00
black.to.1layhid3           5.934973e-02
lstat.to.1layhid3          -6.723087e+00
Intercept.to.1layhid4      -1.263006e+00
crim.to.1layhid4            4.747106e+00
zn.to.1layhid4              2.498340e-01
indus.to.1layhid4          -1.408936e+00
chas.to.1layhid4            3.999629e-01
nox.to.1layhid4             5.050513e-01
rm.to.1layhid4              6.062223e+00
age.to.1layhid4            -9.336642e-01
dis.to.1layhid4             9.443891e-01
rad.to.1layhid4            -1.247986e+00
tax.to.1layhid4            -1.080708e+00
ptratio.to.1layhid4        -4.639221e-01
```

```
black.to.1layhid4      -1.828609e+00
lstat.to.1layhid4      -6.082774e+00
Intercept.to.1layhid5   2.364970e+00
crim.to.1layhid5       -1.154977e+00
zn.to.1layhid5          1.230277e+00
indus.to.1layhid5       1.312696e+00
chas.to.1layhid5        2.739027e-01
nox.to.1layhid5        -1.437402e+00
rm.to.1layhid5         -3.092230e-01
age.to.1layhid5        -3.792320e-01
dis.to.1layhid5        -9.378781e-02
rad.to.1layhid5        -1.985824e-01
tax.to.1layhid5        -1.716352e+00
ptratio.to.1layhid5    -4.595461e-01
black.to.1layhid5      -3.683327e-02
lstat.to.1layhid5      -2.167995e-01
Intercept.to.2layhid1  -1.969399e+00
1layhid1.to.2layhid1    8.118625e-01
1layhid2.to.2layhid1   -8.275659e-02
1layhid3.to.2layhid1    1.673545e+00
1layhid4.to.2layhid1    1.458195e+00
1layhid5.to.2layhid1    1.241880e-01
Intercept.to.2layhid2  -4.952805e-01
1layhid1.to.2layhid2    7.677016e-01
1layhid2.to.2layhid2   -1.712411e+00
1layhid3.to.2layhid2   -3.331395e-01
1layhid4.to.2layhid2   -1.176448e-01
1layhid5.to.2layhid2    2.187176e+00
Intercept.to.2layhid3  -9.683629e-01
1layhid1.to.2layhid3    1.797915e-01
1layhid2.to.2layhid3   -2.016215e+01
1layhid3.to.2layhid3    6.885980e+00
1layhid4.to.2layhid3    5.917666e-01
1layhid5.to.2layhid3   -6.718751e-01
Intercept.to.2layhid4   6.091911e-01
1layhid1.to.2layhid4    7.820839e-01
1layhid2.to.2layhid4   -3.625761e+00
1layhid3.to.2layhid4   -6.641459e+00
1layhid4.to.2layhid4   -2.744620e+00
1layhid5.to.2layhid4    2.604093e+00
Intercept.to.2layhid5  -8.443429e-01
1layhid1.to.2layhid5   -3.574979e-01
1layhid2.to.2layhid5    1.941865e+00
1layhid3.to.2layhid5    6.354234e-01
1layhid4.to.2layhid5   -4.908666e-01
1layhid5.to.2layhid5   -5.367010e-01
Intercept.to.medv      -1.702115e+00
2layhid1.to.medv        1.687521e+00
2layhid2.to.medv        2.562581e+00
2layhid3.to.medv        4.046946e-01
2layhid4.to.medv       -5.766075e-01
2layhid5.to.medv       -1.020548e-01

MSE.net_data
[1] 12.14631

MSE.lm
```

[1] 26.54428

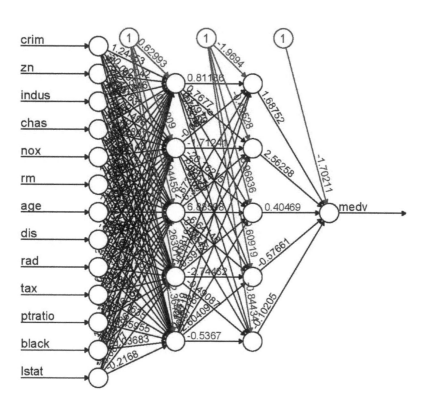

FIGURE 7.5
Neural network with 2 hidden layers of 5 and 5 nodes.

8

Causal Inference and Matching

8.1 Introduction

Randomized controlled study is the golden standard for most of the experiments, which generally requires the participants to be randomly allocated to different groups and gives different treatments (i.e., interventions). For example, in medical research, to test the effect of a treatment or differences between treatments, this setting is usually required by regulatory agencies. To compare the effect differences between different ways of teaching in schools, classes can be divided into cohorts and give the different ways of teaching and compare the effects of learning. With the assumption that randomization does a satisfactory job to make the background covariate distributions of the participants among the groups are balanced (similar but not necessarily identical), one can then attribute the difference of final outcomes to the difference of treatments the participants received. However, in reality, there are as many or more experiments which were not either randomized or controlled, such as data from observational studies, single group of patients with rare disease, or participants cannot be randomized to give different treatments to do certain reasons, it will be more difficult to attribute the final treatment outcomes to one or a few specific factors or causes. That makes this a difficult question to answer for researchers. Over the past decades, with various methodologies proposed by researchers to obtain reasonable solutions, from mathematical to graphical approaches, etc., substantial degrees of successes had been accomplished.

8.2 Three Layer Causal Hierarchy

As proposed by professor Pearl in "The Seven Tools of Causal Inference With Reflections On Machine Learning" [62], there are three layers (or depth) of causal inference problems: (1) Association, (2) Intervention, and (3) Counterfactual. The first level, Association, deals mostly the statistical relationships defined by the observed data and requires no causal information. The second level, Intervention, involves not just seeing what is, but changing what we see

DOI: 10.1201/9781003205685-8

to deal with "What if" problems. The third layer, Counterfactuals, deals with causal effect problem under different conditions and subsumes interventional and associational questions. It has the expressions of the type $P(y_x|x_0, y_0)$, meaning "the probability that event $Y = y$ would be observed had X been x, given that we actually observed $X = x_0$ and $Y = y_0$.

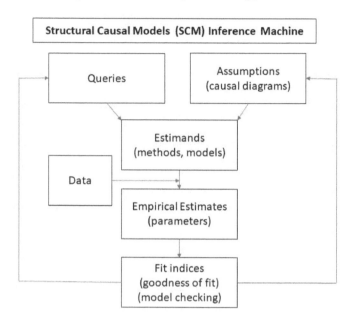

FIGURE 8.1
Causal-effect-SCM.

8.3   Seven Tools of Causal Inference

Pearl and colleagues [63] had proposed a framework which dubbed as Structure Casual Models (SCM), which combines data with the causal model and assumptions via the causal diagram to produce answers to queries of interest. A modified version of the SCM scheme is shown in Figure 8.1. Based on SCM, they described seven essential tools to analyze the problems in the three layers as stated above.

The seven tools are as follows: (1) Encoding Causal Assumptions – Transparency and Testability: Transparency enables analysts to discern whether the assumptions are plausible, or whether additional assumptions are warranted. Testability permits analysts to determine whether the assumptions are compatible with the available data. The use of graphical causal models can mirror the way researchers perceive of cause-effect relationship. Testability is facilitated through a graphical criterion called *d*-separation, which provides the fundamental connection between causes and probabilities. (2) Do-calculus and the control of confounding: Confounding is a major obstacle to drawing causal inference from data, however, this can greatly be illuminated through graphical models via the selection of appropriate set of covariates to control confounding. (3) The Algorithmization of Counterfactuals: Counterfactual analysis tries to estimate the effect if the subject were to take a different course of action. By constructing the causal graphs, one can better formalize counterfactual reasoning by utilizing the scientific knowledge, and to formulate the structural equations and models to estimate the counterfactual effects. (4) Mediation Analysis and the Assessment of Direct and Indirect Effects: Mediation analysis concerns the mechanisms that transmit changes from a cause to its effects. The identification of such intermediate mechanism to define direct and indirect effects is essential for generating explanations and counterfactual analysis. (5) Adaptability, External Validity, and Sample Selection Bias: The representativeness of data sample, the generalizability of the models and validity of the results are all critical and need to be well considered in the course of data analysis and model construction. (6) Recovering from Missing Data: missing data is almost unavoidable in analyzing real data that can potentially increase the reliability of the findings. Well-designed methodologies of missing data imputation need to be incorporated into the processes. (7) Causal Discovery: The *d*-separation criterion and the mechanisms described above enables analysts to detect and enumerate the testable implications of a given causal model and perform inference.

Petersen and van der Laan [64] later elaborated these tools in a more statistical-oriented languages as: specify knowledge about the system to be studied using a causal model, specify the observed data and their link to the causal model, specify the target causal quantity, assess identifiability, commit to a statistical model and estimand, estimate, and interpret the findings.

These tools serve as a reference roadmap to help analysts to understand the problems they are dealing with, the capabilities and the limitations of the tools they can utilize, the procedures to perform the estimations, the goodness of the proposed models, and the possibilities to revisit the validity and sufficiency of original queries and assumptions. Even though these are the conventional procedures for data analysis, how to effectively implement these concepts and processes remains a critical capability for the proficient data analysts.

8.4   Statistical Framework of Causal Inferences

Over the years, causal inferences have been studied by a wide range of researchers, especially via the method of matching using propensity scores and varieties of other measures, for example, see the references in Imbens and Rubin [39].

The concept of propensity scores had been thoroughly discussed by Rosenbaum and Rubin [68], as well as by other authors. In the following, we briefly describe a few key points for data analytical purposes.

Let Y_{i1} denote the response to the experimental treatment of subject $i(1 \leq i \leq N)$, and Y_{i0} denote the response to the control treatment of subject i. Let X_i denote the vector of covariates associated with subject i and $T_i = 1(0)$ if subject i receives experimental (control) treatment. The observed outcome for subject i is then $Y_i = T_i Y_{i1} + (1 - T_i) Y_{i0}$.

If the subjects were appropriately randomized between experimental treatment and control groups, then

$$E(Y_{ij}|T_i = 1) = E(Y_{ij}|T_i = 0), \quad j = 0, 1,$$

even though $E(Y_{i0}|T_i = 1)$ of the experimental treatment group and $E(Y_{i1}|T_i = 0)$ of the control group cannot be estimated from the data since each subject can receive only either experimental or control treatment, but not both.

If the data were appropriately randomized, the estimand of the average treatment effect, which can be estimated empirically using the observed data, can be written as

$$\tau = E(Y_{i1}|T_i = 1) - E(Y_{i0}|T_i = 0),$$

which can be further re-expressed as

$$\tau = a_1[E(Y_{i1}|T_i = 1) - E(Y_{i0}|T_i = 1)] + b_1[E(Y_{i1}|T_i = 0) - E(Y_{i0}|T_i = 0)]$$
$$= a_1\tau_1 + b_1\tau_0, \tag{8.1}$$

where a_1, b_1 are both positive and $a_1 + b_1 = 1$, and

$$\tau_1 = [E(Y_{i1}|T_i = 1) - E(Y_{i0}|T_i = 1)] \text{ and } \tau_0 = [E(Y_{i1}|T_i = 0) - E(Y_{i0}|T_i = 0)]$$

being the (unobserved) treatment effects from the experimental treatment and control groups, respectively.

When the covariate imbalance occurs, proper matchings of subjects to better balance covariates are usually recommended in order to obtain a more appropriate estimate of treatment effect. Given covariate X_i and following the results of Rubin [73], one can show that

$$E(Y_{ij}|X_i, T_i = 1) = E(Y_{ij}|X_i, T_i = 0).$$

Therefore, the treatment effect of the experimental treatment group

$$\tau_1 = E_{\{X_i|T_i=1\}}\{E(Y_i|X_i, T_i = 1) - E(Y_i|X_i, T_i = 0)\},$$

where the expectation is taken over $\{X_i|T_i = 1\}$, can be estimated.

8.5 Propensity Score

Define the propensity score as

$$e(X_i) = P(T_i = 1|X_i) = E(I\{T_i = 1\}|X_i),$$

namely, the probability of patient i being assigned to experimental treatment given the covariate. Assume

(i) $0 < P(T_i = \delta_i|X_i) < 1$ and

(ii) $P(T_1 = \delta_1, \cdots, T_N = \delta_N|X_1, \cdots, X_N) = \prod_{i=1}^{N} e(X_i)^{\delta_i}(1 - e(X_i))^{(1-\delta_i)},$

where $\delta_i = 0$ or 1, Rosenbaum and Rubin showed that

$$\tau_1 = E_{\{e(X_i)|T_i=1\}}\{E(Y_i|e(X_i), T_i = 1) - E(Y_i|e(X_i), T_i = 0)|T_i = 1\}, \quad (8.2)$$

where the expectation is taken over $\{e(X_i)|T_i = 1\}$, and τ_0 can be expressed similarly. Therefore, the average treatment effect can be derived from the estimates of τ_1 and τ_0. More details about the propensity score can be found in Rosenbaum [71] in addition to the papers mentioned herein.

Let $X_i = (x_{i1}, x_{i2}, \cdots, x_{ik})'$ and $m \leq k$ be the vector of covariates. A common method to estimate $e(X_i)$ is via the logit function, i.e.,

$$\text{logit}(e(X_i)) = \beta_0 + h_1(\eta_{1i}) + h_2(\eta_{2i}), \quad (8.3)$$

where h_1 and h_2 are known functions, and

$$\eta_{1i} = \sum_{r=1}^{m} f_r(x_{ir}) \quad \text{and} \quad \eta_{2i} = \sum_{r,q=1}^{m} f_r(x_{ir})f_q(x_{iq})$$

represent the main effects and interactions, respectively. The parameters in Equation (8.3) can be estimated using MLE-based methods. Goodness-of-fit can be checked graphically via, e.g., Landwehr et al. [45] or Tsai [89].

According to Rosenbaum and Rubin (1983), it is advantageous to subclassify or match not only on $e(X)$ but for other functions of X as well. In particular, such a refined procedure may be used to obtain estimates of the average treatment effect in a subpopulation defined by the components of X, e.g., gender or different disease classifications. In addition, in deriving the propensity score, the selection of covariates to be included in the model is critical. It is not necessary true that more covariates are going to produce a better propensity score. However, it is important to include the covariates, which are highly relevant to the questions the analysts try to answer.

8.6   Methodologies of Matching

With the statistical framework and propensity score as a foundation, researchers have been proposing various algorithms to match the subjects between and treatment and control groups and to utilize the estimand as shown in equation (8.2) to estimate the treatment effect. In the following, the various methods for commonly used matching will be discussed. As the derivations and computational procedures are different: therefore, it is expected that the estimates from these methods will be somewhat different.

8.6.1   Nearest Neighbor (or greedy) Matching

This is probably the most commonly used method to perform subject matching. With a defined distance measure between two units, the nearest neighbor, or greedy, matching algorithms move down the list of treated subjects from top to bottom, at each step matching a treated subject to the nearest available control, which is then removed from the list of controls available at the next step. The matching is usually conducted with a given ratio of number of control for each treatment unit, and there is no requirement that all the control units will be used. Matchings are made at a given stage without attention to how they affect possibilities for later matchings.

For the matching with a large reservoir of controls, greedy algorithms often do nearly as well as optimal algorithms (see the next section 8.6.1.1) (Rosenbaum and Rubin 1985). But in the absence of an excess of available controls, or with unfortunate orderings of the list of treated subjects, greedy algorithms can do much worse than optimal ones.

8.6.1.1   Example Using Nearest Neighbor Matching

In this example, we show the nearest neighbor matching using `matchit` function from the `matchIt` R package proposed by Ho et al [34]. Multiple algorithms are also available from other software packages as this is almost the the most commonly used method for matching.

```
library(matchIt)

m.out <- matchit(treat ~ age + educ + nodegree + re74 + re75, method = "nearest",
```

```
                        data = lalonde, ratio=2)
print(summary(m.out))

#output
# shows the improvement of variable balance:

Summary of Balance for All Data:
         Means Treated Means Control Std. Mean Diff. Var. Ratio eCDF Mean eCDF Max
distance       0.3650        0.2738         0.8059      0.7157    0.2008   0.3432
age           25.8162       28.0303        -0.3094      0.4400    0.0813   0.1577
educ          10.3459       10.2354         0.0550      0.4959    0.0347   0.1114
nodegree       0.7081        0.5967         0.2450         .      0.1114   0.1114
re74        2095.5737     5619.2365        -0.7211      0.5181    0.2248   0.4470
re75        1532.0553     2466.4844        -0.2903      0.9563    0.1342   0.2876

Summary of Balance for Matched Data:
         Means Treated Means Control Std. Mean Diff. Var. Ratio eCDF Mean eCDF Max Std. Pair Dist.
distance       0.3650        0.3058         0.5237      0.9694    0.1382   0.2730          0.5293
age           25.8162       26.7486        -0.1303      0.4743    0.0695   0.1378          1.2922
educ          10.3459       10.1838         0.0807      0.4997    0.0353   0.0973          1.1318
nodegree       0.7081        0.6297         0.1724         .      0.0784   0.0784          0.8144
re74        2095.5737     3783.5242        -0.3454      0.9285    0.1437   0.4054          0.5187
re75        1532.0553     2161.4323        -0.1955      1.1738    0.1053   0.2568          0.7299

Percent Balance Improvement:
         Std. Mean Diff. Var. Ratio eCDF Mean eCDF Max
distance           35.0        90.7       31.1     20.5
age                57.9         9.2       14.5     12.6
educ              -46.7         1.1       -1.6     12.6
nodegree           29.6          .        29.6     29.6
re74               52.1        88.7       36.1      9.3
re75               32.6      -258.6       21.5     10.7

# this table shows the number of control matched, unmatched, and discarded:
Sample Sizes:
          Control Treated
All          429     185
Matched      370     185
Unmatched     59       0
Discarded      0       0

# one can also examine how the propensity score was generated:
print(summary(m.out$model))

# output:
Call: glm(formula = treat ~ age + educ + nodegree + re74 + re75,
    family = structure(list(family = "quasibinomial", link = "logit",
    linkfun = function (mu)

Deviance Residuals:
    Min      1Q   Median       3Q      Max
-1.2559  -0.9053  -0.6053   1.2060   2.9809

Coefficients:
              Estimate Std. Error t value Pr(>|t|)
(Intercept) -2.694e+00  8.545e-01  -3.153  0.00170
age          2.464e-03  1.096e-02   0.225  0.82225
educ         1.569e-01  5.668e-02   2.769  0.00580
nodegree     8.502e-01  3.008e-01   2.826  0.00487
re74        -1.225e-04  2.755e-05  -4.446 1.04e-05
re75         2.574e-05  4.231e-05   0.608  0.54320

(Dispersion parameter for quasi-binomial family taken to be 1.144126)
Null deviance: 751.49  on 613  degrees of freedom
Residual deviance: 692.88  on 608  degrees of freedom
Number of Fisher Scoring iterations: 5
```

```
# to see the index of each matched subject between treated and control groups,
# each treated subject was matched to 2 controls (the 2 columns)
print(m.out$match.matrix)

#output
        [,1]      [,2]
NSW1    "PSID382" "PSID424"
NSW2    "PSID261" "PSID102"
NSW3    "PSID151" "PSID159"
NSW4    "PSID410" "PSID235"
NSW5    "PSID350" "PSID119"
...
NSW180  "PSID118" "PSID106"
NSW181  "PSID201" "PSID3"
NSW182  "PSID296" "PSID23"
NSW183  "PSID181" "PSID55"
NSW184  "PSID306" "PSID30"
NSW185  "PSID338" "PSID36"

# one can also use the following commend to get complete list of the control subjects
# with the original data matrix:
print(get_matches(m.out))
```

8.6.2   Exact Matching

This algorithm matches treated individuals to controls with exactly the same values for the covariates of interest. This is more straightforward and does not need to generate propensity scores. But it is probably the most restrictive method to match and mostly used when the covariates are either discrete or discretized continuous variables. Clearly, this creates a good match but power decreases as it usually results in the exclusion of many observations. One can see from the following example output, it only matched about 20 subjects and left lots of subjects not matched.

8.6.2.1   Example

```
library(matchIt)

m.out <- matchit(treat ~ age + educ + nodegree + re74 + re75, method = "exact",
                 data = lalonde)
print(summary(m.out))

#output
# shows the improvement of variable balance:
```

Summary of Balance for All Data:

	Means Treated	Means Control	Std. Mean Diff.	Var. Ratio	eCDF Mean	eCDF Max
age	25.8162	28.0303	-0.3094	0.4400	0.0813	0.1577
educ	10.3459	10.2354	0.0550	0.4959	0.0347	0.1114
nodegree	0.7081	0.5967	0.2450	.	0.1114	0.1114
re74	2095.5737	5619.2365	-0.7211	0.5181	0.2248	0.4470
re75	1532.0553	2466.4844	-0.2903	0.9563	0.1342	0.2876

Summary of Balance for Matched Data:

	Means Treated	Means Control	Std. Mean Diff.	Var. Ratio	eCDF Mean	eCDF Max	Std. Pair Dist.
age	22.5455	22.5455	0	0.983	0	0	0
educ	10.6364	10.6364	0	0.983	0	0	0
nodegree	0.6818	0.6818	-0	.	0	0	0
re74	0.0000	0.0000	0	.	0	0	0
re75	0.0000	0.0000	0	.	0	0	0

```
Percent Balance Improvement:
          Std. Mean Diff. Var. Ratio eCDF Mean eCDF Max
age                   100         97.9      100      100
educ                  100         97.6      100      100
nodegree              100         .         100      100
re74                  100         .         100      100
re75                  100         .         100      100

# this table shows the number of control matched, unmatched, and discarded:
Sample Sizes:
                Control Treated
All              429.       185
Matched (ESS)     16.22      22
Matched           21.        22
Unmatched        408.       163
Discarded          0.         0

# one can also use the following commend to get complete list of the control subjects
# with the original data matrix:
print(match.data(m.out))
```

8.6.3   Mahalanobis Distance Matching

As the covariates are usually multi-dimensional, Mahalanobis distance has been used in multivariate analysis setting to measure the "distance" between two sets of covariates. Specifically, given two covariates, X_i and X_j, the distances between them used in Mahalanobis matching are defined as

$$md(X_i, X_j) = \{(X_i - X_j)'S^{-1}(X_i - X_j)\}^{1/2}, \qquad (8.4)$$

where $S^{1/2}$ is the Cholesky decomposition of the covariance matrix of X of the samples.

8.6.3.1   Example

```
library(Matching)

# to generate the propensity model

data(lalonde)
glm1 <- glm(treat~age + I(age^2) + educ + I(educ^2) + married + nodegree + re74 + I(re74^2) +
            re75 + I(re75^2), family=binomial, data=lalonde)

X <- glm1$fitted
Y <- lalonde$re78
Tr <- lalonde$treat

rr <- Match(Y = Y, Tr = Tr, X = glm1$fitted, estimand = "ATT", M = 1, ties = TRUE, replace = TRUE)
print(summary(rr))

# output
Estimate...  1334.3   # this is the treatment effect
AI SE......  1153.3   # this is the corresponding standard error
T-stat.....  1.157
p.val......  0.24726

Original number of observations.............. 614
Original number of treated obs............... 185
Matched number of observations............... 185
```

```
Matched number of observations  (unweighted).  267

-----------------------------------------------------------------------------
# to check the covariate balance:

mb <- MatchBalance(Tr ~ age + I(age^2) + educ + I(educ^2) + married + nodegree + re74 +
                        I(re74^2) + re75 + I(re75^2) + I(re74 * re75) + I(age * nodegree) +
                        I(educ * re74) + I(educ * re75), match.out = rr, nboots = 1000,
                        data = lalonde)
print(mb)

# mb produces the comparison between before and after matching statistics for every input variable.
# To save space, only one variable (age) is shown here. By comparing the T-test p-values or
# KS bootstrap p-value, one can see the distribution of age was highly different between the
# groups, but the difference become not significant after matching.

***** (V1) age *****
                            Before Matching            After Matching
mean treatment........         25.816                     25.816
mean control..........         28.03                      26.292
std mean diff.........        -30.945                     -6.6544
...
var ratio (Tr/Co).....          0.44                       1.0139
T-test p-value........        0.0029143                   0.43662
KS Bootstrap p-value..      < 2.22e-16                    0.007
KS Naive p-value......        0.0032205                  0.0067145
KS Statistic..........        0.15773                     0.14607


Before Matching Minimum p.value: < 2.22e-16
Variable Name(s): age I(age^2) married re74 I(re74^2) re75 I(re75^2) I(re74 * re75)
                      I(age * nodegree) I(educ * re74) I(educ * re75)

After Matching Minimum p.value: 0.003
Variable Name(s): re74 I(re74^2) I(educ * re74)  Number(s): 7 8 13

# one can also see how the imbalance of variable re74 get improved using graphs.
qqplot(lalonde$re74[rr$index.control], lalonde$re74[rr$index.treated])
abline(coef = c(0, 1), col = 2)
```

8.6.4   Genetic Matching

To increase the flexibility of the Mahalanobis distance, Sekhon [75] introduces, a weight matrix in equation (8.4) and defines a genetic matching algorithm as follows:

$$gmd(X_i, X_j) = \{(X_i - X_j)'S^{-1/2}WS^{-1/2}(X_i - X_j)\}^{1/2}, \qquad (8.5)$$

where W is a diagonal positive definite weight matrix. The elements of W can be chosen objectively to simultaneously minimize the distributional difference and location difference of covariates between the experimental treatment and control groups based on the Kolmogorov-Smirnov test and t-test, respectively. On the other hand, W can also be chosen somewhat subjectively depending on the relative importance among the matched variables. When certain variables are considered as more important, and higher degree of balance for the selected variables is desired, one can assign higher weights for those variables during the matching processes.

The matching can be performed either with pair-matching or full-matching (see the following section) depending on the distributions of the data. The treatment effect can then be estimated between the matched data in the control and experimental groups.

The conventional test of covariate balance between groups based on the *t*-test focuses only on location and can miss distributional differences between the covariates. On the other hand, the Kolmogorov-Smirnov test compares distributional differences and can miss differences in locations. By combining these two tests, matching can often be better assessed.

8.6.4.1 Example

```
library(Matching)

xdat<-lalonde
X<-lalonde[which(colnames(lalonde) %in% c("age","educ","married","nodegree","re74","re75"))]
BalanceMatrix <- cbind(xdat$age,I(xdat$age^2),xdat$educ,I(xdat$educ^2),xdat$married,
                      xdat$nodegree, xdat$re74,I(xdat$re74^2),xdat$re75, I(xdat$re75^2),
                      I(xdat$re74 * xdat$re75), I(xdat$age * xdat$nodegree),
                      I(xdat$educ * xdat$re74), I(xdat$educ * xdat$re75))
gen1 <- GenMatch(Tr = Tr, X = X, BalanceMatrix = BalanceMatrix, pop.size = 1000)

# In order to obtain balance statistics, one can run the following command with the output
# object (gen1) returned by the call to GenMatch above:
mgen1 <- Match(Y = Y, Tr = Tr, X = X, estimand="ATE", Weight.matrix = gen1)
print(summary(mgen1))

#output
Estimate... -830.06
AI SE...... 781.16
T-stat..... -1.0626
p.val...... 0.28796

Original number of observations.............. 614
Original number of treated obs............... 185
Matched number of observations............... 614
Matched number of observations (unweighted). 744

genmb<-MatchBalance(Tr ~ age + I(age^2) + educ + I(educ^2) + married + nodegree + re74 +
                I(re74^2) + re75 + I(re75^2) + I(re74 * re75) + I(age * nodegree) +
                I(educ * re74) + I(educ * re75), data = lalonde, match.out = mgen1,
                nboots = 1000)
print(genmb)

#output to indicate the minimum p-values (i.e., the worst unbalance) during before and
after match:
Before Matching Minimum p.value: < 2.22e-16
Variable Name(s): married re74 I(re74^2) re75 I(re75^2) I(re74 * re75) I(age * nodegree)
                  I(educ * re74) I(educ * re75)

After Matching Minimum p.value: < 2.22e-16
Variable Name(s): re74 I(re74^2) re75 I(re75^2) I(educ * re74) I(educ * re75)

# As for our propensity score examples, the balance output for nodegree, re74 and re74^2
# are presented for close examination:
genmb1<-MatchBalance(Tr ~ nodegree + re74 + I(re74^2), match.out = mgen1,
                nboots = 1000, data = lalonde)
print(genmb1)

***** (V1) nodegree *****
```

	Before Matching	After Matching
mean treatment........	0.70811	0.63029
mean control.........	0.59674	0.63029
std mean diff........	24.431	0
var ratio (Tr/Co).....	0.86157	1

```
T-test p-value........ 0.0069822                    1

***** (V2) re74 *****
                        Before Matching         After Matching
mean treatment........      2095.6                4082.5
mean control.........       5619.2                4555.3
std mean diff........      -72.108               -7.5928
var ratio (Tr/Co).....      0.51813               0.94323
T-test p-value........ 1.7479e-12             4.7201e-08
KS Bootstrap p-value.. < 2.22e-16             < 2.22e-16
KS Naive p-value...... < 2.22e-16             3.1943e-11
KS Statistic..........      0.44704               0.1828

***** (V3) I(re74^2) *****
                        Before Matching         After Matching
mean treatment........     28141412              55373297
mean control.........      77555527              61786421
std mean diff........      -43.306               -5.3367
var ratio (Tr/Co).....      0.65483               0.87015
T-test p-value........ 6.2865e-06             0.01003
KS Bootstrap p-value.. < 2.22e-16             < 2.22e-16
KS Naive p-value...... < 2.22e-16             3.1943e-11
KS Statistic..........      0.44704               0.1828

Before Matching Minimum p.value: < 2.22e-16
Variable Name(s): re74 I(re74^2)

After Matching Minimum p.value: < 2.22e-16
Variable Name(s): re74 I(re74^2)
```

Compare with the propensity score matching before this section, one can see a few points: (1) genetic matching was able to find more matching, (2) the matching balance was very much improved (details not shown here), (3) the average treatment effect was not as much as that estimated by the propensity score matching. Therefore, analysts should try various matching methods to compare the outcomes if feasible.

8.7   Optimal Matching

Optimal matching is originally a sequence analysis method used in social science to assess the dissimilarity of ordered arrays of tokens that usually represent a time-ordered sequence of socio-economic states two individuals have experienced. Once proper distance metrics have been defined and calculated for a set of observations, the commonly used statistical tools can be applied for further data analysis. Rosenbaum [69] had considered the sequence analysis as the matching sequence of a reference matching such as nearest neighbor matching and modify the matching sequence by inserting, deleting, or both to improve the matching results, for example the total difference between the treated and controlled subjects and named it "optimal matching." More specifically, for a sequence of matching,

$$(m_1, m_2, m_3, m_4, \cdots, m_{t-2}, m_{t-1}, m_t),$$

where t is the number of subjects in the treated group, one can insert a new match and create a new match sequence,

$$(m_1, m_2, m^{new}, m_3, m_4, \cdots, m_{t-2}, m_{t-1}, m_t)$$

or delete an existing match

$$(m_1, m_2, m_4, \cdots, m_{t-2}, m_{t-1}, m_t)$$

or create and delete an existing match

$$(m_1, m_2, m^{new}, m_3, m_4, \cdots, m_{t-2}, m_t)$$

so that the total difference will be reduced, therefore, improve the matching. Embedded in the Network Flow theory, the author was able to construct the matching processes effectively, which is either as good or better than the nearest neighbor matching. In addition, with the implementation of caliper, optimal matching can avoid matching the control subjects, which are far away from the treated subjects. One can also use this to minimize total differences that balance the discrete factors. It can also be used to match multiple control subjects.

8.7.0.1 Example

```
library(matchIt)

m.out <- matchit(treat ~ age + educ + nodegree + re74 + re75, method = "optimal",
                 data = lalonde, ratio=1)
print(summary(m.out))

#output
# shows the improvement of variable balance:
```

```
Summary of Balance for All Data:
          Means Treated Means Control Std. Mean Diff. Var. Ratio eCDF Mean eCDF Max
distance      0.3650       0.2738          0.8059       0.7157     0.2008   0.3432
age          25.8162      28.0303         -0.3094       0.4400     0.0813   0.1577
educ         10.3459      10.2354          0.0550       0.4959     0.0347   0.1114
nodegree      0.7081       0.5967          0.2450            .     0.1114   0.1114
re74       2095.5737    5619.2365         -0.7211       0.5181     0.2248   0.4470
re75       1532.0553    2466.4844         -0.2903       0.9563     0.1342   0.2876
```

```
Summary of Balance for Matched Data:
          Means Treated Means Control Std. Mean Diff. Var. Ratio eCDF Mean eCDF Max Std. Pair Dist.
distance      0.3650       0.3626          0.0213       1.0374     0.0088   0.0865      0.0383
age          25.8162      25.4000          0.0582       0.4860     0.0820   0.2378      1.2292
educ         10.3459      10.2919          0.0269       0.5517     0.0273   0.0649      0.9302
nodegree      0.7081       0.7189         -0.0238            .     0.0108   0.0108      0.5945
re74       2095.5737    2105.1528         -0.0020       1.2820     0.0357   0.2216      0.2524
re75       1532.0553    1329.4258          0.0629       2.0898     0.0394   0.1784      0.6212
```

```
Percent Balance Improvement:
          Std. Mean Diff. Var. Ratio eCDF Mean eCDF Max
distance       97.4           89.0        95.6     74.8
age            81.2           12.1        -0.8    -50.8
educ           51.1           15.2        21.3     41.8
nodegree       90.3              .        90.3     90.3
re74           99.7           62.2        84.1     50.4
re75           78.3        -1549.2        70.6     38.0
```

```
# this table shows the number of control matched, unmatched, and discarded:
Sample Sizes:
          Control Treated
All          429     185
Matched      185     185
Unmatched    244       0
Discarded      0       0


# one can also examine how the propensity score was generated:
print(summary(m.out$model))

# output:
Note: this is exactly the same as the propensity scores generated in the nearest method as
      they used the same set of covariates.

# to see the index of each matched subject between treated and control groups,
print(m.out$match.matrix)
#output
          [,1]
NSW1    "PSID368"
NSW2    "PSID396"
NSW3    "PSID293"
NSW4    "PSID227"
NSW5    "PSID394"
...
NSW180 "PSID111"
NSW181 "PSID201"
NSW182 "PSID296"
NSW183 "PSID181"
NSW184 "PSID313"
NSW185 "PSID31"
```

8.8 Full Matching

Given two disjoint sets of $T = \{t_i | i = 1, 2, \cdots, N_T\}$ of treated units and $C = \{c_i | i = 1, 2, \cdots, N_C\}$ of untreated or control units, there are $N_T \times N_C$ pairs (t, c) of possible comparisons. However, it is possible that not all such potential comparisons are reasonable as subjects may differ markedly with respect to pretreatment characteristics or covariates.

For each such pair (t_i, c_j), one can define a non-negative metric $\delta_{(t,c)}$ for similarity or "distance" between the components. Distance can be defined in some broad varieties of similarity measure such as the Euclidean distance, correlation, or multivariate measure such as Mahalanobis distance, etc.

Rosenbuam [70] defined a matching method called "Full matching" for a more flexible and better matching method than the nearest neighbor match in terms of total distance among all the matched pairs. Specifically, he introduced the concept of sub-classification, which assembles treated and control units which are comparable with small distances. Note, a sub-classification may or may not include all the unites in treated and control groups.

Mathematically, a sub-classification with S sub-classes is a collection of S non-empty, mutually exclusive subsets of treated group $(T_1, T_2, \cdots, T_S)$, and a collection of S non-empty, mutually exclusive subsets of control group, $(C_1, C_2, \cdots, C_S)$. The treated units in T_s will be compared with the control units in C_s, and the pair (T_s, C_s) forms a sth subclass, $1 \leq s \leq S$.

Let $N = |T_1 \cup T_2 \cup \cdots, \cup T_S| \leq |T|$ be the total number of treated units in these sub-classes; similarly, let $M = |C_1 \cup C_2 \cup \cdots, \cup C_S| \leq |C|$ be the total number of control units in these sub-classes. Note that if a study has no classifications, then $S = 1$. If $|T_s| = |C_s| = 1$ for all s, then the sub-classification is called a pair-matched sample. If $|T_s| = 1$ for all s, then the sub-classification is called a matched sample with multiple controls. If minimum of $|T_s| = 1$ and $|C_s| = 1$ for all s, then the sub-classification is called a full matching

8.8.0.1 Example

```
library(matchIt)

m.out <- matchit(treat ~ age + educ + nodegree + re74 + re75, method = "full",
                 data = lalonde, ratio=1)
print(summary(m.out))

#output
# shows the improvement of variable balance:
```

Summary of Balance for All Data:

	Means Treated	Means Control	Std. Mean Diff.	Var. Ratio	eCDF Mean	eCDF Max
distance	0.3650	0.2738	0.8059	0.7157	0.2008	0.3432
age	25.8162	28.0303	-0.3094	0.4400	0.0813	0.1577
educ	10.3459	10.2354	0.0550	0.4959	0.0347	0.1114
nodegree	0.7081	0.5967	0.2450	.	0.1114	0.1114
re74	2095.5737	5619.2365	-0.7211	0.5181	0.2248	0.4470
re75	1532.0553	2466.4844	-0.2903	0.9563	0.1342	0.2876

Summary of Balance for Matched Data:

	Means Treated	Means Control	Std. Mean Diff.	Var. Ratio	eCDF Mean	eCDF Max	Std. Pair Dist.
distance	0.3650	0.3651	-0.0008	1.0012	0.0032	0.0378	0.0222
age	25.8162	24.8801	0.1308	0.4963	0.0814	0.2833	1.2769
educ	10.3459	10.3263	0.0098	0.7055	0.0161	0.0402	1.1368
nodegree	0.7081	0.7375	-0.0648	.	0.0294	0.0294	0.8170
re74	2095.5737	2199.3044	-0.0212	1.2023	0.0317	0.2078	0.3741
re75	1532.0553	1387.2168	0.0450	2.0056	0.0473	0.2265	0.8238

Percent Balance Improvement:

	Std. Mean Diff.	Var. Ratio	eCDF Mean	eCDF Max
distance	99.9	99.7	98.4	89.0
age	57.7	14.7	-0.1	-79.6
educ	82.2	50.3	53.7	63.9
nodegree	73.6	.	73.6	73.6
re74	97.1	72.0	85.9	53.5
re75	84.5	-1457.2	64.8	21.3

```
# this table shows the number of control matched, unmatched, and discarded:
```

Sample Sizes:

	Control	Treated
All	429.	185
Matched (ESS)	127.23	185
Matched	429.	185
Unmatched	0.	0
Discarded	0.	0

```
# one can also examine how the propensity score was generated:
```

```
print(summary(m.out$model))

# output:
Note: this is exactly the same as the propensity scores generated in the nearest method as
      they used the same set of covariates.

# to see the index of each matched subject between treated and control groups,
print(m.out$match.matrix)

# one can also examine the various sub-classes and the membership.
# they can also be extracted for further analysis.
```

8.8.1   Analysis of Data After Matching

Matching is just the first step for conducting causal-effect analysis. The treatment-control matched pairs can be considered as "pseudo-randomized" data as in the regular randomization in the randomized-controlled trials. However, one has to be mindful about the weights associated with the pairs created by matching, especially for the full matching algorithm.

The weights created by matching should be used (e.g., in a weighted regression) to ensure that the matched treated and control groups are weighted up to be similar. Analysts should also be aware that the weights created by matching estimate the average treatment effect on the treated, with the control units weighted to resemble the treated units. With subclassification such as in full matching, estimates should be obtained within each subclass and then aggregated across subclasses.

There are two primary ways to estimate treatment effects after doing matching: fixed-effects regression and weighting. The fixed-effects regression explicitly estimates an effect for each matched set. In the case of full matching, these effects are averaged to obtain an overall effect, specifically as proposed by Hansen [30], the effect of the treatment difference between the matched units i can be estimated as follows:

$$Y_i = \begin{cases} \tau_{s(i)} + \Delta_{s(i)} + \epsilon_i & i \in T, \\ \Delta_{s(i)} + \epsilon_i & i \in C \end{cases} \tag{8.6}$$

with $E(\epsilon) = 0, Cov(\epsilon) = \sigma^2 I$ for $\sigma > 0$, and $\Delta_{s(1)}, \Delta_{s(2)}, \cdots, \Delta_{s(S)}$ are matched-set effects and $\tau_{s(1)}, \tau_{s(2)}, \cdots, \tau_{s(S)}$ are treatment control contrasts, one for each matched set. Under model (8.6), in the sth matched set, the average difference of treatment- and control-subgroup responses, $\bar{y}_{st} - \bar{y}_{sc}$, is unbiased estimate τ_s, and the overall treatment effect can be estimated by weighted averages of these differences. Stuart and Green [83] had used this method to estimate the relationship between adolescent marijuana use and adult outcomes.

8.8.1.1 Example

In this section, we will show how the various treatment effects can be estimated using nearest neighbor, optimal, and full matching algorithms. One can realize that the treatment effect estimates are not always identical to different algorithm; therefore, it is imperative to have a follow-up effort to understand the reasons behind the differences and to help the interpretation of the findings.

```
\library(matchIt)

# estimate the Average Treatment Effect on the Treated (ATET) using nearest neighbor algorithm:
m.out1 <- matchit(treat ~ age + educ + nodegree +married + re74 + re75, method = "nearest",
   data = lalonde)

# using least squares model on the control group only.
z.out1 <- zelig(re78 ~ age + educ + nodegree + married + re74 + re75,
   data = match.data(m.out1, "control"), model = "ls")

# set the coefficients of the control groups from above, and apply it to the treated group to get
   the counterfactual effect of the treatment group:
x.out1 <- setx(z.out1, data = match.data(m.out1, "treat"), cond = TRUE)

# conduct simulation of the counterfactual effect of the treated group:
s.out1 <- sim(z.out1, x = x.out1)

# summarize the results:
print(summary(s.out1))

# outputs: the expected (E(Y1|X,T=1)) & predicted (counterfactual) (E(Y1|X,T=0)) treatment effects:
ev
      mean       sd      50%     2.5%    97.5%
1 5553.414 412.4563 5553.454 4759.039 6356.74
pv
         mean      sd      50%     2.5%    97.5%
[1,] 5449.357 5626.94 5392.377 -5501.517 16323.04


# ATET = 5553.414 - 5449.357

--------------------------------------------------------------------
# estimate the Average treatment effect (Oerall) using nearest neighbor algorithm:
m.out1 <- matchit(treat ~ age + educ + nodegree +married + re74 + re75, method = "nearest",
    data = lalonde)

# continue with the above and fit the linear model to the treatment group:
z.out2 <- zelig(re78 ~ age + educ + nodegree + married + re74 + re75,
   data = match.data(m.out1, "treat"), model = "ls")

# conduct the same simulation procedure in order to impute the counterfactual outcome
   for the control group,
x.out2 <- setx(z.out2, data = match.data(m.out1, "control"), cond = TRUE)
s.out2 <- sim(z.out2, x = x.out2)
print(summary(s.out2))

# outputs: the expected (E(Y0|X,T=0)) & predicted (counterfactual) (E(Y0|X,T=1)) control effects:
ev
      mean       sd      50%     2.5%    97.5%
1 6352.133 590.2209 6347.349 5176.554 7504.396
pv
         mean      sd      50%     2.5%    97.5%
[1,] 6036.209 7938.089 6017.632 -9576.08 20755.45

# Overall effect =
Average  of counterfactual treatment  and counterfactual control effect
```

```
= [(5553.414 - 5449.357) + (6036.209 - 6352.133)] /2 = -105.9335
```

In this part, we show the same estimate for Average Treatment Effect on the Treated (ATET) and Overall treatment effects using Optimal and Full matching, one can easily see the difference between these estimates, due to the difference in matching algorithms. Optimal and Full matching use more matching than the nearest neighbor, namely, they use more data; and therefore, the results tend to be more complete than those from the nearest neighbor matching.

```
\library(matchIt)

# estimate the Average Treatment Effect on the Treated (ATET) using Optimal matching algorithm:
m.out1 <- matchit(treat ~ age + educ + nodegree +married + re74 + re75, method = "optimal",
    data = lalonde)

# summarize the results:
print(summary(s.out1))

# outputs: the expected (E(Y1|X,T=1)) & predicted (counterfactual) (E(Y1|X,T=0)) treatment effects:
ev
        mean        sd      50%      2.5%     97.5%
1 5517.433 419.7703 5508.129 4735.889 6369.076
pv
        mean        sd      50%      2.5%     97.5%
[1,] 5568.073 5734.803 5533.061 -5703.013 16585.72

# ATET = 5517.433 - 5568.073

------------------------------------------------------------------
# estimate the Average treatment effect (Overall) using Optimal matching algorithm
# outputs: the expected (E(Y0|X,T=0)) & predicted (counterfactual) (E(Y0|X,T=1)) control effects:
ev
        mean        sd      50%      2.5%     97.5%
1 6342.141 576.9575 6333.528 5219.135 7500.566
pv
        mean        sd      50%      2.5%     97.5%
[1,] 6548.983 8059.922 6637.913 -8497.92 22658.24

# Overall effect =
Average  of counterfactual treatment  and counterfactual control effect
= [(5517.433 - 5568.073) + (6548.983 - 6342.141)] /2 = 78.101

------------------------------------------------------------------

# estimate the Average Treatment Effect on the Treated (ATET) using Full matching algorithm:
m.out1 <- matchit(treat ~ age + educ + nodegree +married + re74 + re75, method = "optimal",
    data = lalonde)

# summarize the results:
print(summary(s.out1))

# outputs: the expected (E(Y1|X,T=1)) & predicted (counterfactual) (E(Y1|X,T=0)) treatment effects:
ev
        mean        sd      50%    2.5%    97.5%
1 6986.387 307.4181 6993.52 6359.3 7588.946
pv
        mean        sd      50%      2.5%     97.5%
[1,] 7024.151 6356.641 7215.337 -5307.594 19193.66

# ATET = 6986.387 - 7024.151
```

```
--------------------------------------------------------------------
# estimate the Average treatment effect (Overall) using Full matching algorithm
# outputs: the expected (E(YO|X,T=0)) & predicted (counterfactual) (E(YO|X,T=1)) control effects:
ev
      mean       sd       50%     2.5%    97.5%
1 6378.908 576.0216 6364.997 5277.356 7547.425
pv
      mean       sd       50%     2.5%    97.5%
[1,] 6488.295 7980.147 6419.71 -9247.214 21602.42

# Overall effect =
Average  of counterfactual treatment  and counterfactual control effect
= [(6986.387 - 7024.151) + (6488.295 - 6378.908)] /2 = 35.812
```

8.9   Cluster Matching

In medical research, it is a common practice to enroll subjects for a new experimental treatment in multiple nations so that the new medicines can be used for all subjects who may need it. Similarly, in social studies, the survey may be conducted across several regions in a country or over the world. Due to the potential differences in culture, medical practices, lifestyles, etc., across the regions, the experimental treatment effect may be different; therefore, in the estimation of causal effects for the treatment, it is essential to take these differences into consideration. Arpino and Cannas [5] proposed a method to perform the subject matching within the regions before combining into an estimate of the overall effect, which is named as cluster matching.

Specifically, instead of matching the treated with control subjects in a whole group, they proposed a two-level matching structure, with the following approaches: (1) Single-level propensity score model, matching on the whole data set from all regions as the usual practices. (2) Single-level propensity score model, within-cluster matching, which find the matched subjects within the same cluster. (3) Single-level propensity score model, preferential within-cluster matching, which matches within the cluster first, with the option to find matched subjects in other clusters. (4) Fixed-effects propensity score model, matching on the pooled data set using fixed effects models. (5) Random-effects propensity score model, which matches on the pooled data set using random effects models.

The matching starts with the construction of propensity score using all data from all clusters.

$$\text{logit}(e_{ij}) = \alpha_0 + X_{ij}\beta. \tag{8.7}$$

Then, a one-to-one caliper matching (could be any matching methods discussed in the previous sections) on the estimated propensity scores is implemented on the pooled data set.

Let S_t and S_c denote the set of treated and control units, respectively, let $A_{crj} \in S_c$ indicate the set of control units matched to $A_{trj} \in S_t$, the rth treated subject for cluster j. Then all treated units that successfully found a matched control unit and all matched control units are included in data set, M, while all other units are discarded:

$$M = \{A_{trj} \mid A_{crj} \neq \phi\} \cup \{A_{crj} \mid \forall \, rj\}$$

and the average treatment effect for the treated (ATET) can be estimated by

$$ATET = \frac{1}{card(M)} \left\{ \sum_{rj \in S_t \cap M} \{Y_{rj} - \sum_{kl \in S_c} Y_{kl} \cdot w(rj, kl)\} \right\} \qquad (8.8)$$

Approach (1) is a simple single-level logit model and is the same as what had been discussed above.

The second approach, within-cluster matching, with the estimated propensity scores as in equation (8.7), the matching is implemented within clusters, and the ATET is estimated using equation (8.8) with $l = j$ in the summation. One disadvantage of within cluster matching is that the search for the match is only conducted within the same cluster and can possibly miss better matches in other clusters and may end up with having too many subjects been discarded.

The third approach, preferential within-cluster matching, starts by searching control units within the same cluster, if none is found to meet the matching criteria, control units are searched in other clusters. This approach improves the balancing and potentially reduces the number of unmatched treated units comparing with the within-cluster matching only.

The fixed-effects and random-effects models take a regression approach by specifying the cluster-level variables, the nested subject-level variables, and the treatment indicator. Similar to equation (8.7), one can specify the regression model to include the selected variables, even including the propensity score, and not limited to the variables for generating propensity score. One can then utilize the linear or non-linear mixed effects model to model the treatment effect individually by subjects. This approach is different from the usual matching method; however, it could offer other advantages from regression setting.

8.9.1   Example

In the following, we demonstrate how to use cluster matching to estimate the ATET of a large data set from the National Education Longitudinal Study (https://nces.ed.gov/surveys/nels88/). This was a nationally representative, longitudinal study of 8th graders in 1988. Sample of students followed throughout secondary and post-secondary years. Surveys of students reporting on school, work, home experiences, educational resources and support, the role in education of parents and peers, neighborhood characteristics, educational and occupational aspirations, and other student perceptions. Additional topics included self-reports on smoking, alcohol, and drug use and extracurricular activities. Student assessments in reading, social studies, mathematics, and science. Students' teachers, parents, and school administrators were also surveyed. The data set analyzed is only a subsample of this big data set. Using propensity scores and multivariate data, various matching algorithms, including within-school (cluster), preferentially within-school, random effects models, etc., are used.

```
library(CMatching)
library(lme4)

# define the response and input variable matrix:
data(schools)
X<-schools$ses      # input matrix
Y<-schools$math     # response variable
Tr<-ifelse(schools$homework>1,1,0)    # homework?
Group<-schools$schid    # school id.

# Multivariate Matching on covariates in X, one-to-one matching on X with replacement.
# Matching within schools
mw<-CMatch(type="within", Y=Y, Tr=Tr, X=X, Group=Group, caliper=0.1)

# compare balance before and after matching
bmw <- CMatchBalance(Tr~X,data=schools,match.out=mw)

# output: balance of covariates are substantially improved: (e.g., T-test)
***** (V1) X *****
                          Before Matching        After Matching
mean treatment........    0.23211                0.25282
mean control..........    -0.36947               0.25541
std mean diff.........    61.315                 -0.26998
var ratio (Tr/Co).....    1.2849                 1.0154
T-test p-value........    3.4478e-07             0.61904
KS Bootstrap p-value..    < 2.22e-16             1
KS Naive p-value......    1.979e-07              0.99907
KS Statistic..........    0.35227                0.055556

# examine summary output:
print(summary(mw))
#output:
Original number of observations..............  260
Original number of treated obs...............  128
Matched number of observations...............  85
Matched number of observations  (unweighted).  90
Caliper (SDs)........................................  0.1
Number of obs dropped by 'exact' or 'caliper' .........  43
Estimate... 4.3412    # (ATET)
SE........  1.8323    # (se for ATET)

-------------------------------------------------------------------------

# Match preferentially within school
mpw <- CMatch(type="pwithin",Y=schools$math, Tr=Tr, X=schools$ses, Group=schools$schid, caliper=0.1)
# examine covariate balance
bmpw<- CMatchBalance(Tr~ses,data=schools,match.out=mpw)

# outut:
***** (V1) ses *****
                          Before Matching        After Matching
mean treatment........    0.23211                0.23211
mean control..........    -0.36947               0.23148
std mean diff.........    61.315                 0.063703
var ratio (Tr/Co).....    1.2849                 1.0055
T-test p-value........    3.4478e-07             0.89232
KS Bootstrap p-value..    < 2.22e-16             0.972
KS Naive p-value......    1.979e-07              0.98207
```

```
KS Statistic.........    0.35227                0.054054

# see summary output
print(summary(mpw))

#output:
Original number of observations.............   260
Original number of treated obs..............   128
Matched number of observations..............   128
Matched number of observations (unweighted).  148
Caliper (SDs).......................................   0.1
Number of obs dropped by 'exact' or 'caliper' ........   0
Estimate...  5.5078
SE........   2.2046

-------------------------------------------------------------------
# Propensity score (ps) matching
# estimate the ps model
mod <- glm(Tr~ses+parented+public+sex+race+urban,family=binomial(link="logit"),data=schools)
eps <- fitted(mod)

# within school propensity score matching
psmw <- CMatch(type="within",Y=schools$math, Tr=Tr, X=eps, Group=schools$schid, caliper=0.1)
print(summary(psmw))

# output:
Original number of observations.............   260
Original number of treated obs..............   128
Matched number of observations..............   95
Matched number of observations (unweighted).  95
Caliper (SDs).......................................   0.1
Number of obs dropped by 'exact' or 'caliper' ........   33
Estimate...  3.9895
SE........   2.0519

-------------------------------------------------------------------
# preferential within school propensity score matching
psmw <- CMatch(type="pwithin",Y=schools$math, Tr=Tr, X=eps, Group=schools$schid, caliper=0.1)
print(summary(psmw))

# output:
Original number of observations.............   260
Original number of treated obs..............   128
Matched number of observations..............   128
Matched number of observations (unweighted).  129
Caliper (SDs).......................................   0.1
Number of obs dropped by 'exact' or 'caliper' ........   0
Estimate...  2.6914
SE........   2.3509

-------------------------------------------------------------------
# propensity score matching using ps estimated from multilevel logit model
# (random intercept at the hospital level)
# require(lme4)
mod<-glmer(Tr ~ ses + parented + public + sex + race + urban + (1 | schid),
        family=binomial(link="logit"), data=schools)
eps <- fitted(mod)
mpsm<-CMatch(type="within",Y=schools$math, Tr=Tr, X=eps, Group=NULL, caliper=0.1)
print(summary(mpsm))

# output:
Original number of observations.............   260
Original number of treated obs...............   128
Matched number of observations..............   123
Matched number of observations (unweighted).  133
Number of obs dropped by 'exact' or 'caliper'  5
Estimate...  2.3902
```

```
AI SE......  1.4997
T-stat.....  1.5938
p.val......  0.11098
```

There are a few notable points with different matching methods: (1) using multivariate data, the preferentially within-school matching shows greater effect on the mathematics scores than the within-school only matching, but it comes with larger variation of the estimate. (2) When using propensity scores, the effects seem to be reversed with within-school only matching having a larger effect. (3) The random effect model shows similar results with the preferentially within-school matching using propensity score. (4) Using multivariate data, the ATET effect is statistically significant at the 5% level; however, none of the results based on propensity score show significance at this level.

9

Business and Commercial Data Modeling

Up to this chapter, the examples shown are mostly based on big data from the field of medical research. In this chapter, big data from field of commercial business will be used to illustrate the applicability of the methodologies discussed previously. The data bases are included in UCI Machine Learning Repository (https://archive.ics.uci.edu/ml/datasets.php).

9.1 Case Study One: Marketing Campaigns of a Portuguese Banking Institution

9.1.1 Description of Data

The first data are related to direct marketing campaigns of a Portuguese banking institution. The marketing campaigns were based on phone calls. Often, more than one contact to the same client was required, in order to access if the product (bank term deposit) would be ("yes") or not ("no") subscribed. It contains 45211 records with 17 attributes of client information and 4 national social economic index indicators. The goal is to predict if the client will subscribe (yes/no) a term deposit (variable y). The attribute (input variable) information are as follows:

```
[1]  "1. age (numeric)"
[2]  "2. job : type of job"
[3]  "3. marital : marital status"
[4]  "4. education"
[5]  "5. default: has credit in default?"
[6]  "6. housing: has housing loan?"
[7]  "7. loan: has personal loan?"
[8]  "8. contact: contact communication type"
[9]  "9. month: last contact month of year"
[10] "10. day of week: last contact day of the week"
[11] "11. duration: last contact duration, in seconds"
[12] "12. campaign: number of contacts for this client"
[13] "13. pdays: number of days passed from a previous campaign"
[14] "14. previous: number of contacts before this campaign"
[15] "15. poutcome: outcome of the previous marketing campaign"
[16] "16. emp.var.rate: employment variation rate - quarterly indicator"
[17] "17. cons.price.idx: consumer price index - monthly indicator"
[18] "18. cons.conf.idx: consumer confidence index - monthly indicator"
[19] "19. euribor3m: euribor 3 month rate - daily indicator"
[20] "20. nr.employed: number of employees - quarterly indicator"
```

DOI: 10.1201/9781003205685-9

```
[21] "21. Output variable (desired target): y - has a term deposit?"
```

9.1.2   Data Analysis

The various methods discussed in the previous chapters are used to analyze
the bank marketing data. The brief R-codes and results are showing below.
The relevant plots are also shown.

9.1.2.1   Analysis via Lasso

```
fit<-glmnet(x, y, family=c("binomial"), alpha = 1, nlambda = 100,
            data=BankNew.dat)
cvfit<-cv.glmnet(x, y, family=c("binomial"))
plot(cvfit)
bestlam <- cvfit$lambda.min
print(bestlam)

[1] Best lambda after CV: 0.003232541

bestcoef<-coef(cvfit)
print("Non-zero coefficients from minimum lambda of cross-validation")

#outputs:
                      [,1]
(Intercept) -5.339982e+02
marital      6.092330e-02
duration     4.960214e-03
euribor3m    1.086991e+02
```

As shown above, lasso selected the marital status, duration, and the eu-
ribor3m as the more influential variables for determining the marketing outcomes. The graph from cross-validation to determine the best number of variables (3) for this analysis is shown below:

9.1.2.2   Analysis via Elastic Net

Following shows the outputs from **enet**:

```
# elasticNet
Call: enet(x = x, y = y, lambda = smincvcv[1], max.steps = 50)

# output:
Sequence of  moves: (of variables selection)
   duration  nr.employed  pdays  emp.var.rate  cons.conf.idx  contact
   euribor3m  education  month job  cons.price.idx  marital  age
   previous day_of_week  loan  campaign  poutcome  housing  default
```

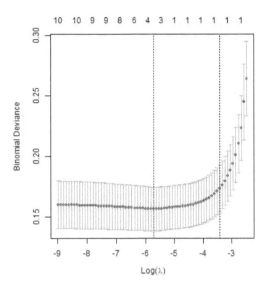

FIGURE 9.1
Important variable selected by lasso.

The output from **enet** shows the sequence of variable selection, which indicates the relative importance of these variables in terms of predicting the outcomes.

9.1.2.3 Analysis via SIS

The following shows the results of data analysis using ISI methods.

```
# using: SIS(x, y, family='binomial', tune='cv', penalty=c(SCAD))
# output - variables selected:
(Intercept)      X1      X2      X3      X4      X6      X7      X8
    -60.111   0.004   0.042   0.065  -0.104  -0.013  -0.078   0.915
             X9     X10     X11     X12     X13     X14     X15     X16
         -0.068   0.010   0.004  -0.035  -0.002  -0.194   0.270  -0.883
            X17     X18     X19     X20
          0.986   0.042   0.320  -0.007

# using: SIS(x, y, family='binomial', tune='aic', penalty=c(SCAD))
# output - variables selected:
(Intercept)      X2      X3      X4      X8      X9     X11     X13
     69.096   0.010   0.002  -0.056   0.677  -0.008   0.004  -0.002
            X14     X18     X20
         -0.344   0.033  -0.014
```

```
# using: SIS(x, y, family='binomial', tune='bic', varISIS='vanilla',
          penalty=c(MCP))
# output - variables selected:
(Intercept)    X2     X4     X8     X9    X11    X13    X14
   -105.502  0.041 -0.103  0.893 -0.034  0.004 -0.002 -0.322
         X16    X17    X18    X20
      -0.826  1.226  0.056 -0.002
```

Data analysis using SIS needs to take into consideration of the various options of **tune** and penalty choices. As shown in this example, different choices can lead to different results, and which results is making more sense needs to consult with the subject-matter experts.

9.1.2.4 Analysis via rpart

The following shows the results of data analysis using the **rpart** method.

```
fit1 <- rpart(y ~ . , data=BankNew.dat)
print(summary(fit1))
rpart.plot(fit1)

# outputs:
n= 41188
node), split, n, deviance, yval
      * denotes terminal node

 1) root 41188 4117.2850 0.11265420
   2) nr.employed>=5087.65 36224 2267.8550 0.06711020
     4) duration< 525.5 32317  821.0101 0.02608534
       8) month>=4.5 29918  418.9627 0.01420550 *
       9) month< 4.5 2399  345.1680 0.17423930 *
     5) duration>=525.5 3907  942.5575 0.40645000
      10) duration< 835.5 2536  547.6341 0.31545740 *
      11) duration>=835.5 1371  335.0868 0.57476290 *
   3) nr.employed< 5087.65 4964 1225.9860 0.44500400
     6) duration< 172.5 1872  269.8798 0.17467950 *
     7) duration>=172.5 3092  736.4877 0.60866750
      14) pdays>=16.5 2192  546.7664 0.52372260 *
      15) pdays< 16.5 900  135.3822 0.81555560 *
```

The codes also requested the tree to be created and it is shown in Figure 9.2

9.1.2.5 Analysis via randomForest

The results are shown in the following using **randomForest**:

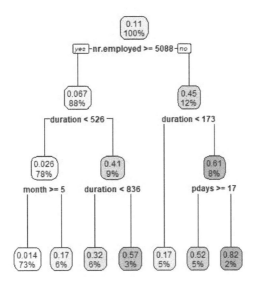

FIGURE 9.2
Recursive partitioning of variables by **rpart**.

```
randomForest(formula = factor(y) ~ ., data = BankNew.dat,
importance = TRUE, proximity = TRUE, na.action = na.omit)

Outputs:
Type of random forest: classification
Number of trees: 500
No. of variables tried at each split: 4
OOB estimate of  error rate: 2.91%
Confusion matrix:
      0 1 class.error
0 3029 1 0.000330033
1   90 2 0.978260870
```

The confusion matrix indicates that the prediction was not as good for the category of "1 (=yes)" as the number of marketing success was not many. This is not unusual for prediction in this case if the number of success/failure is very different, which is the case here. The graph for the important variables from randomForest for this analysis is shown Figure 9.3: it shows the variable **duration** as the most important variable follows by education, age, job, and marital status, etc.

bank.rf

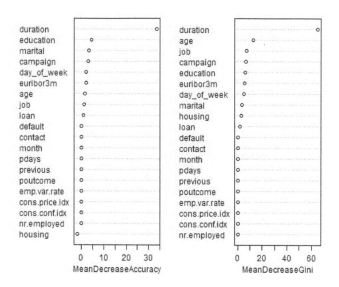

FIGURE 9.3
Important variable selected by randomForest.

9.1.2.6   Analysis via xgboost

This section shows the analysis using xgboost:

```
dtrain<-xgb.DMatrix(x,label=y)
param <- list(max_depth = 4, eta = 1, verbose = 0, nthread = 2,
objective = "binary:logistic")
watchlist <- list(train = dtrain, eval = dtrain)
bst <- xgb.train(param, dtrain, nrounds = 2, watchlist)
print(bst)
print(xgb.importance(model = bst))
xgb.plot.importance(xgb.importance(model = bst))

Outputs:
evaluation_log:
 iter train_error eval_error
    1      0.0246     0.0246
    2      0.0226     0.0226

       Feature        Gain        Cover  Frequency
1:    duration 0.875363460 0.919678605 0.57894737
2:     marital 0.028820459 0.003300013 0.05263158
```

```
3:  day_of_week  0.026474222  0.005495470  0.05263158
4:          loan  0.022215931  0.021473801  0.05263158
5:      campaign  0.014208550  0.012287623  0.05263158
6:           job  0.013243280  0.009967638  0.05263158
7:      euribor3m  0.012326820  0.018173787  0.05263158
8:           age  0.007347279  0.009623063  0.10526316
```

The outputs show the low degree of misclassification and eight important variables including duration and marital status as the most influential predictors. Following that, other variables are also predictive as lasso and randomForest had shown. The importance of the variables is also shown graphically in the following:

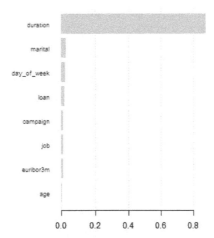

FIGURE 9.4
Important variable selected by xgboost.

9.2   Summary

There are quite a few notable differences between the results derived from different statistical methods used:

1. All of these methods selected 3. marital : marital status and

11. duration: last contact duration, in seconds as the first two most important variables for the success of marketing campaign followed by other variables in various orders. Analysts need to understand the reasonable rationale why that is the case. Of course, one cannot know the precise reasons, but good understanding of the background will help to fine-tune the model or future success of campaigns.

2. The results from some methods can be very much depending on the various options provided in the program, which indicates the various assumptions in the theory or computations. Default choices may not always be the best to use as every data set has its own nature. Analysts need to take into considerations of the options.

3. Given the nature of the difficulties of this type of promotional campaign, the successful outcomes are much less than failures, and most of the modeling will produces high degree of mis-classification error (e.g., as shown in `randomForest`). Even though `xgboost` seems to produce much better results, it still needs to be validated with other data sets with similar natures.

4. The graphical displays from tree-based modeling usually are quite useful to convey the findings and discuss with the users; however, the tree should not include too many levels as that will be more difficult for the users to absorb the findings. The balance between tree size and details of the findings needs to be well considered.

9.3 Case Study Two: Polish Companies Bankruptcy Data

9.3.1 Description of Data

The data set was created by Sebastian Tomczak of Department of Operations Research, Wrocław University of Science and Technology, Poland. It is about bankruptcy prediction of Polish companies. The data were collected from Emerging Markets Information Service, which is a database containing information on emerging markets around the world. The bankrupt companies were analyzed in the period 2000–2012, while the still operating companies were evaluated from 2007 to 2013. Based on the collected data, five classification cases were distinguished that depends on the forecasting period. The total number of companies analyzed, bankrupted, or not-bankrupted for each specific year is shown in the following table.

Data_Year	Number_of_Companies	Number_bankrupted	Number_not_bankrupted
Year-1	7027	271	6756
Year-2	10173	400	9773
Year-3	10503	495	10008

Year-4	7792	515	9277
Year-5	5910	410	5500

The data collected included various metrics regarding a company operations, accounting metrics, profitability, etc. For the companies which data were collected during Year-1, the status whether the company bankrupted after 5 years was recorded. Similarly, for the companies which data were collected during Year-2, 3, 4, 5, the status whether the company bankrupted after 4, 3, 2, 1 year(s) was recorded. The goal of the following analysis is to predict the company bankruptcies based on these financial metrics.

The financial metrics collected are shown as follows:

```
"X1 net profit/total assets"
"X2 total liabilities/total assets"
"X3 working capital/total assets"
"X4 current assets/short-term liabilities"
"X5 [(cash+short-term securities+receivables-short-term liabilities)
    /(operating expenses-depreciation)]*365"
"X6 retained earnings/total assets"
"X7 EBIT/total assets"
"X8 book value of equity/total liabilities"
"X9 sales/total assets"
"X10 equity/total assets"
"X11 (gross profit+extraordinary items+financial expenses)/total assets"
"X12 gross profit/short-term liabilities"
"X13 (gross profit+depreciation)/sales"
"X14 (gross profit+interest)/total assets"
"X15 (total liabilities*365)/(gross profit+depreciation)"
"X16 (gross profit+depreciation)/total liabilities"
"X17 total assets/total liabilities"
"X18 gross profit/total assets"
"X19 gross profit/sales"
"X20 (inventory*365)/sales"
"X21 sales(n)/sales(n-1)"
"X22 profit on operating activities/total assets"
"X23 net profit/sales"
"X24 gross profit (in 3 years)/total assets"
"X25 (equity-share capital)/total assets"
"X26 (net profit+depreciation)/total liabilities"
"X27 profit on operating activities/financial expenses"
"X28 working capital/fixed assets"
"X29 logarithm of total assets"
"X30 (total liabilities-cash)/sales"
"X31 (gross profit+interest)/sales"
"X32 (current liabilities*365)/cost of products sold"
"X33 operating expenses/short-term liabilities"
"X34 operating expenses/total liabilities"
"X35 profit on sales/total assets"
"X36 total sales/total assets"
"X37 (current assets-inventories)/long-term liabilities"
"X38 constant capital/total assets"
"X39 profit on sales/sales"
"X40 (current assets-inventory-receivables)/short-term liabilities"
"X41 total liabilities/((profit on operating activities+depreciation)*(12/365))"
"X42 profit on operating activities/sales"
"X43 rotation receivables+inventory turnover in days"
"X44 (receivables*365)/sales"
"X45 net profit/inventory"
"X46 (current assets-inventory)/short-term liabilities"
"X47 (inventory*365)/cost of products sold"
"X48 EBITDA (profit on operating activities-depreciation)/total assets"
"X49 EBITDA (profit on operating activities-depreciation)/sales"
```

```
"X50 current assets/total liabilities"
"X51 short-term liabilities/total assets"
"X52 (short-term liabilities*365)/cost of products sold)"
"X53 equity/fixed assets"
"X54 constant capital/fixed assets"
"X55 working capital"
"X56 (sales-cost of products sold)/sales"
"X57 (current assets-inventory-short-term liabilities)/(sales-gross profit-depreciation)"
"X58 total costs/total sales"
"X59 long-term liabilities/equity"
"X60 sales/inventory"
"X61 sales/receivables"
"X62 (short-term liabilities *365)/sales"
"X63 sales/short-term liabilities"
"X64 sales/fixed assets"
```

9.3.2   Data Analysis

As a first step of data analysis, one needs examine data quality. The data have a substantial number of outliers for each variable and they were adjusted using the robust `winsorize` function in R package `DescTools`, which bring the data outside of the `prob=`$(0.1\%, 99.9\%)$ back to the respective percentiles. The results are plotted in Figure 9.5. The left panel has a few very small number (x-variable < -400) and a few larger number. The right panel shows the data distribution after the adjustment.

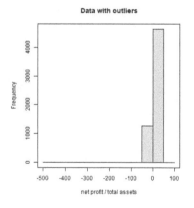 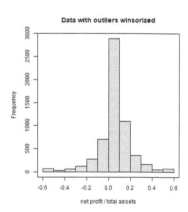

FIGURE 9.5
Examine data quality to winsorize outliers.

In addition, preliminary data examination also shows many metrics that were highly correlated. The following shows the metric-pairs with correlations > 0.8 by liability, assets, other operational measures.

	Financial-Metric-1	Financial-Metric-2	Correlation
1	X1 net profit/tot_assets	X7 EBIT/tot_assets	0.965
2	X1 net profit/tot_assets	X11 (gr_profit+extra_items+fin exp.)/tot_assets	0.931

3	X1 net profit/tot_assets	X14 (gr_profit+interest)/tot_assets	0.965
4	X1 net profit/tot_assets	X18 gr_profit/tot_assets	0.961
5	X1 net profit/tot_assets	X22 profit on oper act/tot_assets	0.872
6	X4 cur_assets/st_liab	X46 (cur_assets-inven)/st_liab	0.917
7	X7 EBIT/tot_assets	X11 (gr_profit+extra_items+fin exp.)/tot_assets	0.964
8	X7 EBIT/tot_assets	X14 (gr_profit+interest)/tot_assets	1.000
9	X7 EBIT/tot_assets	X18 gr_profit/tot_assets	0.996
10	X7 EBIT/tot_assets	X22 profit on oper act/tot_assets	0.917
11	X8 book value of equity/tot_liab	X17 tot_assets/tot_liab	0.986
12	X8 book value of equity/tot_liab	X50 cur_assets/tot_liab	0.853
13	X10 equity/tot_liab	X25 (equity-share capital)/tot_liab	0.811
14	X11 (gr_profit+extra_items+fin exp.)/tot_assets	X14 (gr_profit+interest)/tot_assets	0.964
15	X11 (gr_profit+extra_items+fin exp.)/tot_assets	X18 gr_profit/tot_assets	0.960
16	X11 (gr_profit+extra_items+fin exp.)/tot_assets	X22 profit on oper act/tot_assets	0.915
17	X13 (gr_profit+depr)/sales	X19 gr_profit/sales	0.937
18	X13 (gr_profit+depr)/sales	X23 net profit/sales	0.827
19	X13 (gr_profit+depr)/sales	X31 (gr_profit+interest)/sales	0.934
20	X13 (gr_profit+depr)/sales	X42 profit on oper act/sales	0.810
21	X14 (gr_profit+interest)/tot_assets	X18 gr_profit/tot_assets	0.997
22	X14 (gr_profit+interest)/tot_assets	X22 profit on oper act/tot_assets	0.917
23	X16 (gr_profit+depr)/tot_liab	X26 (net profit+depr)/tot_liab	0.994
24	X17 tot_assets/tot_liab	X50 cur_assets/tot_liab	0.846
25	X18 gr_profit/tot_assets	X22 profit on oper act/tot_assets	0.912
26	X19 gr_profit/sales	X23 net profit/sales	0.881
27	X19 gr_profit/sales	X31 (gr_profit+interest)/sales	0.990
28	X19 gr_profit/sales	X42 profit on oper act/sales	0.871
29	X19 gr_profit/sales	X49 (profit on oper act-depr)/sales	0.839
30	X22 profit on oper act/tot_assets	X35 profit on sales/tot_assets	0.865
31	X23 net profit/sales	X31 (gr_profit+interest)/sales	0.872
32	X23 net profit/sales	X42 profit on oper act/sales	0.931
33	X23 net profit/sales	X49 (profit on oper act-depr)/sales	0.884
34	X28 working capital/fixed assets	X54 constant capital/fixed assets	0.990
35	X31 (gr_profit+interest)/sales	X42 profit on oper act/sales	0.847
36	X31 (gr_profit+interest)/sales	X49 (profit on oper act-depr)/sales	0.814
37	X32 (current liabilities*365)/sales	X52 (st_liab*365)/cops	0.957
38	X33 oper_exp./st_liab	X63 sales/st_liab	0.973
39	X39 profit on sales/st_liab	X42 profit on oper act/sales	0.853
40	X40 (cur_assets-inven-recv)/st_liab	X46 (cur_assets-inven)/st_liab	0.835
41	X42 profit on oper act/sales	X49 (profit on oper act-depr)/sales	0.932

During the data analysis, one either deletes the highly correlated variables before the analysis or let the statistical methods select the variables (call it univariate analysis) or combine them with proper statistical weighting (call it composite analysis).

The analytical results based on recursive partition (randomForest & xgboost), regularized shrinkage (lasso & SIS), and support vector machine (svm, ksvm, sigFeature) are shown below. In general, none of the methods predict bankruptcies very well except for xgboost, which is due to the reason that there are very few bankruptcies for prediction. Only the results from Year-1,3,5 are shown here as other years also show similar conclusion. The results from the univariate analysis are presented below followed by that of the composite analysis. The analysis was conducted in the order of randomForest, xgboost, lasso, SIS, svm, ksvm, and sigFeature, and their results are presented below in this order.

9.3.2.1   Analysis of Year-1 Data (univariate analysis)

```
***** randomForest *****

Call: randomForest(formula = as.factor(Y) ~ ., data = xdat2, importance = TRUE,
    proximity = TRUE, na.action = na.omit)
Type of random forest: classification
Number of trees: 500
No. of variables tried at each split: 8
OOB estimate of  error rate: 1.14%
Confusion matrix:
    0 1 class.error
```

```
0  3115  6  0.001922461
1    30  0  1.000000000
```

```
***** xgboost *****
```

```
call: xgb.train(params = param, data = dtrain, nrounds = 2, watchlist = watchlist)
params (as set within xgb.train):
  max_depth = "4", eta = "1", verbose = "0", nthread = "2", objective = "binary:logistic",
  validate_parameters = "TRUE"
```

```
nfeatures : 64
evaluation_log:
 iter train_error eval_error
    1    0.009521   0.009521
    2    0.009521   0.009521
    Feature       Gain         Cover  Frequency
1:     X37 0.264337338 0.0028354781 0.06666667
2:     X46 0.200353164 0.2272256396 0.13333333
3:     X21 0.192622772 0.0010863274 0.06666667
4:     X30 0.091761198 0.0884950936 0.13333333
5:      X9 0.076916899 0.5144428148 0.26666667
6:     X20 0.076679479 0.0008114402 0.06666667
7:     X53 0.049029031 0.0762313761 0.06666667
8:     X15 0.039964674 0.0871226792 0.13333333
9:      X1 0.008335445 0.0017491509 0.06666667
```

```
***** Lasso *****
```

```
[1] "Non-zero coefficients from minimum lambda of cross-validation"
      (Intercept)      X1    X3 X5    X11    X13 X15    X23   X25   X27    X30 X34     X38     X39
[1,]      2.674 -8.518 0.27   0 5.196 -0.71   0 -3.461 0.336 0.001 1.911 0.5 -4.689 -4.121
          X41    X44    X46    X49    X51    X52   X53    X57    X58
[1,] -0.042 0.006 -2.062 6.608 -2.61 -0.513 0.261 -0.021 -2.906
          X59    X61    X62    X63    X64
[1,] -0.008 -0.021 -0.007 -0.164 -0.014
```

```
***** SIS   *****
```

```
[1] "model21$coef.est"
      no variable selected
```

```
*****  SVM   *****
```

```
Call: svm(formula = y ~ x)
Parameters:
 SVM-Type:  eps-regression
 SVM-Kernel: radial
 cost:  1
 gamma:  0.015625
 epsilon:  0.1
Number of Support Vectors: 195
```

```
Call: ksvm(formula = y ~ x)
SV type: eps-svr  (regression)
parameter : epsilon = 0.1  cost C = 1
Gaussian Radial Basis kernel function.
 Hyperparameter : sigma =  0.0289
Number of Support Vectors : 257
Training error : 0.964429
```

The important variables from **randomForest** and **xgboost** are shown in Figure 9.6.

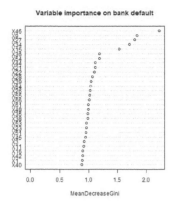 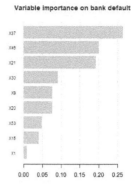

FIGURE 9.6
Variable importance from randomForest and xgboost (Year-1).

9.3.2.2 Analysis of Year-3 Data (univariate analysis)

```
***** randomForest *****

Call: randomForest(formula = as.factor(Y) ~ ., data = xdat2, importance = TRUE,
      proximity = TRUE, na.action = na.omit)
Type of random forest: classification
Number of trees: 500
No. of variables tried at each split: 8
OOB estimate of  error rate: 2.23\%
Confusion matrix:
     0 1 class.error
0 4723 6 0.001268767
1  106 0 1.000000000

***** xgboost *****

call: xgb.train(params = param, data = dtrain, nrounds = 2, watchlist = watchlist)
params (as set within xgb.train):
  max_depth = "4", eta = "1", verbose = "0", nthread = "2", objective = "binary:logistic",
  validate_parameters = "TRUE"

nfeatures : 64
evaluation_log:
 iter train_error eval_error
    1    0.021096    0.021096
    2    0.020683    0.020683
    Feature        Gain        Cover  Frequency
 1:     X34 0.18723330 0.1101329044 0.14285714
 2:     X46 0.17445138 0.1769753349 0.19047619
 3:     X26 0.13796788 0.2503298504 0.09523810
 4:     X58 0.12287635 0.0331910123 0.09523810
 5:     X24 0.10818714 0.0417360710 0.04761905
 6:     X39 0.08912172 0.0004538887 0.04761905
 7:      X1 0.08765618 0.1870618084 0.14285714
 8:     X21 0.02826163 0.0003923904 0.04761905
 9:      X6 0.02487995 0.1391959419 0.09523810
10:      X2 0.02269618 0.0124768063 0.04761905
```

```
11:     X9 0.01666829 0.0480539912 0.04761905


***** Lasso *****

[1] "Non-zero coefficients from minimum lambda of cross-validation"
     (Intercept)    X2      X3      X4      X9     X10     X11    X12     X13 X15    X20     X22
[1,]      2.909 0.537 2.325 0.819 -0.311 0.004 -1.336 0.234 -10.09    0 0.008 -0.012
         X23   X24    X25    X26    X28   X29    X30   X31   X32   X33
[1,] 6.074 0.866 -0.675 -2.064 -0.006 0.167 -0.472 3.925 0.006 0.247
         X34   X35 X37   X38    X39   X40    X41    X44    X45    X46    X47   X48 X49
[1,] 0.822 0.022   0 -2.07 -4.316 0.162 -0.056 -0.001 -0.006 -1.203 -0.011 1.815 0.13
         X50   X51    X52   X54 X55   X56    X57   X58   X59 X60
[1,] -1.984 -0.665 -2.486 0.063   0 3.218 -0.017 -3.33 0.008   0
         X61 X62   X63    X64
[1,] -0.032   0 -0.641 -0.011


*****  SIS   *****

[1] "model21$coef.est"
(Intercept)         X25
    -3.697       -0.354


*****  SVM   *****

Call: svm(formula = y ~ x)
Parameters:
 SVM-Type:  eps-regression
 SVM-Kernel: radial
 cost:  1
 gamma:  0.015625
 epsilon:  0.1
Number of Support Vectors:  622


Call: ksvm(formula = y ~ x)
SV type: eps-svr  (regression)
parameter : epsilon = 0.1  cost C = 1
Gaussian Radial Basis kernel function.
 Hyperparameter : sigma = 0.0402
Number of Support Vectors : 878
Training error : 0.93623
```

The important variables from **randomForest** and **xgboost** are shown in Figure 9.7.

9.3.2.3 Analysis of Year-5 Data (univariate analysis)

```
***** randomForest *****

Call: randomForest(formula = as.factor(Y) ~ ., data = xdat2, importance = TRUE,
      proximity = TRUE, na.action = na.omit)
Type of random forest: classification
Number of trees: 500
No. of variables tried at each split: 8
OOB estimate of  error rate: 3.34%
Confusion matrix:
      0 1 class.error
0 2886 8  0.00276434
1   92 9  0.91089109


***** xgboost *****

call: xgb.train(params = param, data = dtrain, nrounds = 2, watchlist = watchlist)
params (as set within xgb.train):
  max_depth = "4", eta = "1", verbose = "0", nthread = "2", objective = "binary:logistic",
```

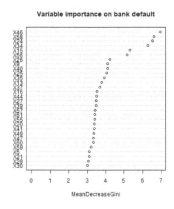

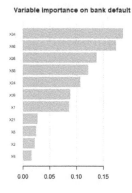

FIGURE 9.7
Variable importance from randomForest and xgboost (Year-3).

```
validate_parameters = "TRUE"

nfeatures : 64
evaluation_log:
 iter train_error eval_error
    1    0.024708    0.024708
    2    0.020367    0.020367
     Feature        Gain        Cover Frequency
 1:      X27 0.226642014 0.1722984240      0.04
 2:      X39 0.106720654 0.2388758380      0.08
 3:      X56 0.105830728 0.0122963605      0.04
 4:      X46 0.102286125 0.0098374058      0.04
 5:      X24 0.069705104 0.0729783516      0.08
 6:      X34 0.065249951 0.0119483744      0.08
 7:      X15 0.048810313 0.0073591787      0.04
 8:      X41 0.039196257 0.1643019362      0.08
 9:      X38 0.033447413 0.0010728242      0.04
10:      X57 0.031615166 0.0014382172      0.04
11:      X40 0.028648724 0.0074787296      0.04
12:      X29 0.027672637 0.0669373615      0.04
13:      X11 0.024542417 0.0013806885      0.04
14:      X37 0.023806891 0.0016270164      0.04
15:      X47 0.020788675 0.0013806885      0.04
16:      X25 0.014584054 0.0005177582      0.04
17:      X43 0.012106408 0.0653103503      0.04
18:      X64 0.010577862 0.1610803296      0.04
19:       X9 0.004264343 0.0005177582      0.04
20:       X3 0.003504265 0.0013624083      0.08


***** Lasso *****

[1] "Non-zero coefficients from minimum lambda of cross-validation"
      (Intercept)     X3 X5     X6      X9    X11     X13 X15      X17    X20     X21    X24
[1,]        -0.09 -0.43  0  0.009 -0.118  1.908 -0.284    0 -0.125  0.001  -1.351  1.515
         X25     X26 X27    X28     X29    X30    X31    X34    X35     X36
[1,] -0.781 -2.936    0  0.008 -0.079  0.087  0.876  0.635  -6.08 -0.263
         X38     X39   X40    X41 X44     X46    X48    X49     X50     X51     X52    X53
[1,] -1.323 -3.467  0.25 -0.022   0 -0.302   1.33  0.171 -0.555 -0.193 -0.155  0.006
        X56     X57     X58    X59 X60    X61    X62    X63     X64
```

```
[1,]   3.249 0.027 -0.291 0.014   0 0.005 0.003 -0.098 0.013
```

```
*****   SIS   *****
```

```
[1] "model21$coef.est"
(Intercept)      X3      X21      X24      X25      X26      X28      X29      X34
   -2.380  -0.415  -0.609   0.935  -0.721  -0.495   0.005  -0.061   0.213
      X35      X38      X46      X62      X64
   -5.670  -0.800  -0.068   0.003   0.005
```

```
*****   SVM   *****
```

```
Call: svm(formula = y ~ x)
Parameters:
 SVM-Type:  eps-regression
 SVM-Kernel:  radial
 cost:  1
 gamma:  0.015625
 epsilon:  0.1
Number of Support Vectors:  515
```

```
Call: ksvm(formula = y ~ x)
SV type: eps-svr  (regression)
parameter : epsilon = 0.1  cost C = 1
Gaussian Radial Basis kernel function.
 Hyperparameter : sigma =  0.0326
Number of Support Vectors : 689
Training error : 0.787563
```

The important variables from **randomForest** and **xgboost** are shown in Figure 9.8.

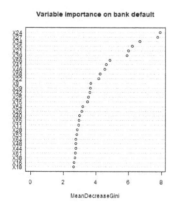
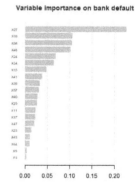

FIGURE 9.8

Variable importance from randomForest and xgboost (Year-5).

9.3.2.4 Analysis of Year-1 Data (composite analysis)

```
***** randomForest *****
```

```
Call: randomForest(formula = as.factor(Y) ~ ., data = xdat2, importance = TRUE,
        proximity = TRUE, na.action = na.omit)
Type of random forest: classification
Number of trees: 500
No. of variables tried at each split: 7
OOB estimate of  error rate: 0.95%
Confusion matrix:
    0 1 class.error
0 3121 0          0
1   30 0          1


***** xgboost *****

call: xgb.train(params = param, data = dtrain, nrounds = 2, watchlist = watchlist)
params (as set within xgb.train):
  max_depth = "4", eta = "1", verbose = "0", nthread = "2", objective = "binary:logistic",
  validate_parameters = "TRUE"

nfeatures : 62
evaluation_log:
  iter train_error eval_error
    1    0.008569    0.008569
    2    0.007299    0.007299
    Feature       Gain         Cover  Frequency
 1:  icaD31 0.124959260 0.0009735568 0.05555556
 2:  icaD32 0.124040971 0.0009555203 0.05555556
 3:  icaD61 0.120220601 0.0067995507 0.05555556
 4:  icaD48 0.104338012 0.0024388932 0.05555556
 5:  icaD57 0.072008783 0.0759151981 0.05555556
 6:  icaD11 0.067372285 0.1746580078 0.05555556
 7:   icaD2 0.062482643 0.0058259938 0.05555556
 8:   icaD1 0.062414694 0.0006972328 0.05555556
 9:   icaD9 0.047108032 0.0005176224 0.05555556
10:  icaD13 0.046861840 0.2368449861 0.11111111
11:  icaD38 0.044131366 0.0067069562 0.05555556
12:  icaD46 0.043463489 0.2426019087 0.11111111
13:  icaD40 0.035827093 0.0684184203 0.05555556
14:  icaD27 0.023404212 0.0048704732 0.05555556
15:  icaD21 0.013361286 0.0042680630 0.05555556
16:  icaD62 0.008005433 0.1675076165 0.05555556


***** Lasso *****

[1] "Non-zero coefficients from minimum lambda of cross-validation"
      (Intercept) icaD2 icaD5 icaD13 icaD16 icaD19 icaD36 icaD40 icaD46
[1,]       -4.966 0.149 0.098  0.304 -0.054 -0.247 -0.002  0.342  0.134
      icaD47 icaD48 icaD57 icaD61 icaD62
[1,]   0.103 -0.048 -0.124  0.172 -0.117


*****   SIS   *****

[1] "model21$coef.est"
(Intercept)         X13         X40
     -4.701       0.126       0.236


*****   SVM   *****

Call: svm(formula = y ~ x)
Parameters:
 SVM-Type:  eps-regression
 SVM-Kernel:  radial
 cost:  1
```

```
gamma:  0.01612903
epsilon:  0.1
Number of Support Vectors:  399

Call: ksvm(formula = y ~ x)
SV type: eps-svr  (regression)
parameter : epsilon = 0.1  cost C = 1
Gaussian Radial Basis kernel function.
 Hyperparameter : sigma =  0.03086
Number of Support Vectors : 567
Training error : 0.901886

Call: svm.default(x = x[, featureRankedList], y = y, kernel = "linear", cost = 10)
Parameters:
 SVM-Type:  eps-regression
 SVM-Kernel:  linear
 cost:  10
 gamma:  0.01612903
 epsilon:  0.1
Number of Support Vectors:  117
```

The important variables from **randomForest** and **xgboost** are shown in Figure 9.9.

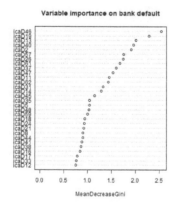 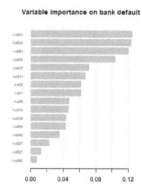

FIGURE 9.9

Variable importance from randomForest and xgboost (Year-1).

9.3.2.5   Analysis of Year-3 Data (composite analysis)

```
***** randomForest *****

Call: randomForest(formula = as.factor(Y) ~ ., data = xdat2, importance = TRUE,
      proximity = TRUE, na.action = na.omit)
Type of random forest: classification
Number of trees: 500
No. of variables tried at each split: 7
OOB estimate of  error rate: 2.13%
Confusion matrix:
```

```
        0 1   class.error
0 4728 1 0.0002114612
1  102 4 0.9622641509
```

```
***** xgboost *****
```

```
call: xgb.train(params = param, data = dtrain, nrounds = 2, watchlist = watchlist)
params (as set within xgb.train):
  max_depth = "4", eta = "1", verbose = "0", nthread = "2", objective = "binary:logistic",
  validate_parameters = "TRUE"
```

```
nfeatures : 63
evaluation_log:
 iter train_error eval_error
    1    0.017373   0.017373
    2    0.013650   0.013650
     Feature       Gain        Cover  Frequency
 1:   icaD27 0.201885106 0.0060588736 0.04347826
 2:   icaD49 0.152625816 0.0048104676 0.04347826
 3:   icaD47 0.116369182 0.2397918661 0.08695652
 4:   icaD31 0.096435382 0.0006213729 0.04347826
 5:   icaD63 0.067087101 0.0018308496 0.04347826
 6:    icaD2 0.059562430 0.2406860678 0.13043478
 7:   icaD33 0.058252168 0.0062431578 0.08695652
 8:   icaD24 0.045853789 0.0803171236 0.08695652
 9:  icaD288 0.039886014 0.0004246854 0.04347826
10:   icaD39 0.037842318 0.0008188488 0.04347826
11:   icaD25 0.036049701 0.0054375006 0.04347826
12:    icaD8 0.031755996 0.0009227734 0.04347826
13:   icaD35 0.024370071 0.0087543692 0.08695652
14:   icaD46 0.022274353 0.2343771935 0.08695652
15:   icaD55 0.006848497 0.1686176581 0.04347826
16:    icaD1 0.002902073 0.0002871921 0.04347826
```

```
***** Lasso *****
```

```
[1] "Non-zero coefficients from minimum lambda of cross-validation"
      (Intercept)  icaD1 icaD2 icaD3 icaD4 icaD6 icaD7 icaD8 icaD9 icaD10 icaD11
[1,]       -4.803 -0.095 0.183 0.243 0.067 0.826 0.346 0.178 0.136  0.104  0.187
      icaD12 icaD13 icaD14 icaD15 icaD16 icaD17 icaD18 icaD20 icaD21 icaD22
[1,]    0.49 -0.129  0.212  0.056   0.03  -0.25 -0.334  -0.06 -0.007  0.027
      icaD23 icaD24 icaD25 icaD26 icaD27 icaD288 icaD29 icaD30 icaD31 icaD32 icaD33
[1,]   0.027 -0.881 -0.023   0.01 -0.327  -0.068  0.102  0.373  -0.37  0.209 -0.237
      icaD34 icaD35 icaD37 icaD38 icaD39 icaD40 icaD41 icaD42 icaD44
[1,]  -0.384  0.226  0.111 -0.732  0.006 -0.201  0.465 -0.136 -0.057
      icaD46 icaD47 icaD49 icaD50 icaD53 icaD54 icaD55 icaD56 icaD57 icaD59 icaD62
[1,]  -0.051  0.374  0.388  0.574 -0.088  0.222  0.094  0.105  0.091  0.077  0.001
```

```
*****   SIS    *****
```

```
[1] "model21$coef.est"
(Intercept)       X47       X49
     -3.817     0.049     0.164
```

```
*****   SVM    *****
```

```
Call: svm(formula = y ~ x)
Parameters:
 SVM-Type:  eps-regression
 SVM-Kernel:  radial
 cost:  1
 gamma:  0.01587302
 epsilon:  0.1
Number of Support Vectors:  747
```

```
Call: ksvm(formula = y ~ x)
SV type: eps-svr  (regression)
```

```
parameter : epsilon = 0.1   cost C = 1
Gaussian Radial Basis kernel function.
 Hyperparameter : sigma =  0.0351
Number of Support Vectors : 1076
Training error : 0.856414

Call: svm.default(x = x[, featureRankedList], y = y, kernel = "linear", cost = 10)
Parameters:
 SVM-Type:  eps-regression
 SVM-Kernel: linear
 cost:  10
 gamma:  0.01587302
 epsilon:  0.1
Number of Support Vectors:  278
```

The important variables from `randomForest` and `xgboost` are shown in Figure 9.10.

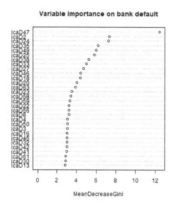 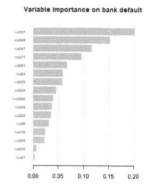

FIGURE 9.10
Variable importance from randomForest and xgboost (Year-3).

9.3.2.6   Analysis of Year-5 Data (composite analysis)

```
***** randomForest *****

Call: randomForest(formula = as.factor(Y) ~ ., data = xdat2, importance = TRUE,
     proximity = TRUE, na.action = na.omit)
Type of random forest: classification
Number of trees: 500
No. of variables tried at each split: 7
OOB estimate of  error rate: 2.94%
Confusion matrix:
      0  1  class.error
0 2893  1 0.0003455425
1   87 14 0.8613861386

***** xgboost *****
```

```
call: xgb.train(params = param, data = dtrain, nrounds = 2, watchlist = watchlist)
params (as set within xgb.train):
  max_depth = "4", eta = "1", verbose = "0", nthread = "2", objective = "binary:logistic",
  validate_parameters = "TRUE"

nfeatures : 63
evaluation_log:
 iter train_error eval_error
    1    0.023706   0.023706
    2    0.020701   0.020701
     Feature       Gain          Cover Frequency
 1:   icaD31 0.13975673 0.1723275120      0.05
 2:   icaD55 0.11145493 0.1709465904      0.05
 3:   icaD33 0.09584604 0.0786970948      0.05
 4:   icaD42 0.06316304 0.0031340463      0.05
 5:   icaD38 0.05458725 0.1694505920      0.05
 6:    icaD4 0.05361662 0.0755630501      0.05
 7:    icaD3 0.05295409 0.0015739862      0.05
 8:   icaD40 0.04976420 0.0009890937      0.05
 9:    icaD6 0.04849602 0.0014959984      0.05
10:   icaD41 0.04636689 0.0013809216      0.05
11:    icaD9 0.04179597 0.0008140151      0.05
12:   icaD56 0.03980501 0.0011729888      0.05
13:   icaD48 0.03886270 0.0009781528      0.05
14:   icaD37 0.03638470 0.0745739542      0.05
15:   icaD22 0.03603275 0.1689902848      0.05
16:   icaD25 0.03142346 0.0729999643      0.05
17:   icaD46 0.02524839 0.0024903896      0.10
18:    icaD7 0.02402531 0.0019610576      0.05
19:    icaD2 0.01041590 0.0004603072      0.05


***** Lasso *****

[1] "Non-zero coefficients from minimum lambda of cross-validation"
     (Intercept)  icaD1 icaD3 icaD4  icaD5  icaD8  icaD9 icaD10 icaD11 icaD12 icaD13 icaD15
[1,]      -4.688 -0.006 -0.47 0.348 -0.179 -0.209 -0.122  0.127  0.098  0.303 -0.348  -0.33
     icaD16 icaD17 icaD18 icaD19 icaD20 icaD21 icaD22 icaD23
[1,] -0.029  0.441 -0.279 -0.078  0.405  0.007 -0.053  0.144
     icaD24 icaD25 icaD26 icaD288 icaD29 icaD30 icaD31 icaD32 icaD33 icaD34 icaD36 icaD37
[1,]  0.374 -0.006  0.202  -0.076  0.021 -0.638 -0.462 -0.135 -0.437  0.552  0.164 -0.291
     icaD38 icaD39 icaD42 icaD43 icaD44 icaD45 icaD47 icaD48
[1,] -0.078  0.178 -0.691  0.147 -0.149   0.05 -0.082 -0.288
     icaD49 icaD50 icaD53 icaD55 icaD56 icaD58 icaD59 icaD60 icaD61 icaD62 icaD63
[1,] -0.125 -0.135 -0.256  0.013 -0.026 -0.069  0.147 -0.138      0  0.399  0.051


*****  SIS  *****

[1] "model21$coef.est"
(Intercept)     X3     X4    X12    X13    X15    X17    X18    X23    X30    X31    X33
     -3.809 -0.276  0.243  0.234 -0.179 -0.206  0.269 -0.116  0.084 -0.270 -0.330 -0.353
        X34    X37    X39    X42    X43    X44    X46    X48    X62
      0.042 -0.206  0.036 -0.084  0.110 -0.026 -0.024 -0.079  0.295


*****  SVM  *****

Call: svm(formula = y ~ x)
Parameters:
 SVM-Type:  eps-regression
 SVM-Kernel:  radial
  cost:  1
 gamma:  0.015873
 epsilon:  0.1
Number of Support Vectors:  660

Call: ksvm(formula = y ~ x)
SV type: eps-svr  (regression)
parameter : epsilon = 0.1  cost C = 1
```

```
Gaussian Radial Basis kernel function.
 Hyperparameter : sigma =  0.0295
Number of Support Vectors : 856
Training error : 0.702214

Call: svm.default(x = x[, featureRankedList], y = y, kernel = "linear", cost = 10)
Parameters:
 SVM-Type:  eps-regression
 SVM-Kernel:  linear
 cost:   10
 gamma:   0.01587302
 epsilon:  0.1
Number of Support Vectors:   266
```

The important variables from **randomForest** and **xgboost** are shown in Figure 9.11.

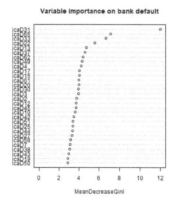 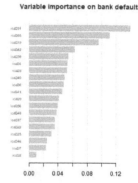

FIGURE 9.11
Variable importance from randomForest and xgboost (Year-5).

9.4   Summary

There are quite a few notable differences between the statistical methods used, the univariate vs composite analytical approach, and between the years:

1. Comparing the results between various methods for the same year, one can notice that the important variables selected by these methods are quite different. Most of the time, they have selected a few the most important variables because these variables are so influential for the prediction, beyond that, the additional variables selected

could be quite different. This happens in both the univariate and composite approaches.

2. Lasso tends to select more variables than any other methods such as `randomForest, xgboost, SIS`, which confirms with the common criticism that lasso is likely to be more liberal in selecting too many variables in the model.

3. The support vector machine methods such as `svm, ksvm` use all the variables by design and tried to select the support vectors to conduct classification. `svm` usually selects more variables than `ksvm`, however that is not always the case, especially in regression problems. `ksvm` usually has lower error rate than that of `svm`. `sigFeature` incorporates variable selection in the estimation, given that, it usually selects fewer support vectors than `svm, ksvm` does and it also reduces the classification error rate.

4. Composite approach performs better than the univariate approach; however, that should not be a surprise as it incorporates the information in all variables which are aggregated by weighting through Independent Component Analysis `ica`. Comparing each year and each method, respectively, composite approach always has smaller error rate than the corresponding analysis using the univariate approach. However, it is more challenging in interpreting the findings from the composite approach. One has to be aware that each component in the `ica` analysis is weighted by the original set of variables. The higher weight means the corresponding variables are more influential for the component.

5. It is usually not easy to predict bank default given the complexities of the operations. The financial metrics are the usually measures to use; however, as shown in this analysis that these metrics are very inadequate for this purpose. Another difficulty is that only a very small number of bankruptcies occurred in the population of banks, which makes most of the statistical methods for this purpose fail to achieve good prediction. Even though methods of sampling can be attempted to improve the prediction but the results are mixed.

6. Even though the original data have five years of bankruptcies data, but given that no bank identification, this is not a longitudinal data for individual bank; therefore, one cannot use the longitudinal data method to better track the performance of a bank. That could be which point one can improve in collecting data for the similar kind of analysis.

10

Analysis of Response Profiles

10.1 Introduction

Majority of clinical trial data analyses focus only on the last response of a patient and paying little attention to the responses before that. However, analysis of the response profile longitudinally can help to understand how the treatment works during the course of the treatments. In medical research with dynamic treatment allocation designs, patients' responses are tracked at every cycle or visit so that the proper treatment can be assigned for the next cycle of the trial. Therefore, it is crucial to extend the conventional data analysis to cover a broader scope so that the analytical results can be more informative to the clinical communities and the patients.

In this chapter, we provide a case study of clinical trial data examining the whole response profile of each patient instead of just the last observation and trying to investigate the potential linkage of the responses with clinical factors and patient's background. Using data from a recent clinical trial, we demonstrate the treatment effects longitudinally by graphical methods, we also estimate the probability of changing disease status during the study, and the similarity of response profiles so that patients can be segmented into sub-groups or clusters. We further investigate the difference between the clusters by examining the baseline data and the data collected during the course of the study. The longitudinal response profiles are also analyzed using a generalized estimating equation to identify the treatment effects as well as the effects of covariates.

Note: even though this research was conducted under the clinical trial setting, the same methodologies and procedures can be applied to other areas of research, including market research, epidemiology, etc., with longitudinal data for behavior tracking or long-term trend.

10.2 Data Example

The following are the response profiles from a sample of patients in a clinical trial with chemotherapy. The responses can be either stable disease (SD),

DOI: 10.1201/9781003205685-10

progressive disease (PD), partial response (PR), nearly complete response (CP), or complete response (CR). The number of treatment cycles ranged from 1 cycle to more than 20 cycles. We only analyze the data from the first 22 cycles due to the sparseness of the patient population beyond that. According to the study protocol, whenever a patient had progression disease, the patient's treatment was terminated.

	pt	trt	rv3	rv5	rv7	rv8	rv9	rv10	rv11	rv12	rv13	rv14	rv15	rv16	rv17	rv18	rv19	rv20	rv21	rv22
1	1014021	A	SD	SD	SD	SD	SD	PR	PR	PR	PR	PR	PR	PR						
2	1014838	A	SD	SD	SD	SD	PR	PR												
4	1025268	B	SD																	
5	1034019	A	NE	NE	SD	PR	PR	PR	PR	PR	PR	PR	PR	PR	PR		PR	PR	PD	PD
6	1034426	B	SD																	
7	1034430	A	SD	SD	PD															
8	1034812	A	PD																	
9	1034815	B	SD	SD	SD	SD	PD	PD												
10	1034822	A	SD	SD	SD	PR	PR	PR	CP	CP	NE	PD								
11	1034831	A	SD	SD	PR	PR	PR	PR	PR	PR	PR	PD								

One can also present the response profiles graphically (Figure 10.1), which gives an easier visualization to examine the responses. One can also display the cumulative responses longitudinally according to the duration of the study (Figure 10.2). As one can easily see that lots of patients with SD status at the beginning of the study, and toward the end of the study, many patients had dropped from the study (green area) and substantially a number of patients had PR, CP, or CR. A graphical comparison between treatments can help to evaluate the efficacy of the treatment, as shown in Figures 10.3 and 10.4.

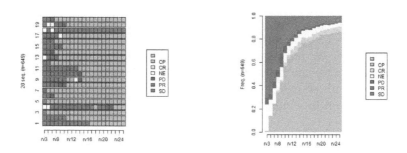

FIGURE 10.1
20 Response Profiles.

FIGURE 10.2
Cumulative Profiles.

Similar to the analysis of discrete data, one can also create the frequency table to demonstrate the frequencies of each type of response profile as shown in the following. For example, the first row below "PD/1-20" indicates the subject stayed in the state of "PD" (progression disease) for 1 cycle and no information for the next 20 cycles and that was because, as soon as the patient

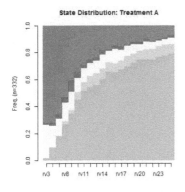

State Distribution: Treatment A

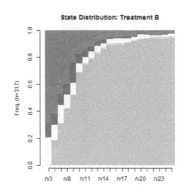

State Distribution: Treatment B

FIGURE 10.3
Group A.

FIGURE 10.4
Group B.

had progression disease, the patient would be taken off the study; therefore, there was no information afterwords. And the row with "(SD/2-PR/2-PD/1-/16)" indicates that the patient had "SD" status for 2 cycles to start with, followed by 2 cycles of "PR" (partial response), then the disease turned worse and progressed for 1 cycle, then he/she was taken off the study until the end of the study; there were 13 cases with this pattern. The disease profiles for all the patients are summarized in the following so that one can easily understand the progress of the disease during the history of the treatment.

```
[>] 656 sequences with 5 distinct events/states including missing value as additional state
[>] 243 distinct sequences. min/max sequence length: 1/21
```

	Freq	Percent
PD/1-/20	37	5.70
SD/1-/20	37	5.70
SD/4-PD/1-/16	30	4.62
SD/2-/19	25	3.85
SD/1-PD/1-/19	24	3.70
SD/2-PD/1-/18	22	3.39
SD/3-PD/1-/17	14	2.16
SD/2-PR/2-PD/1-/16	13	2.00
NE/1-/20	10	1.54
SD/3-/18	10	1.54
SD/5-PD/1-/15	9	1.39
NE/1-SD/1-PD/1-/18	8	1.23
SD/2-PR/1-PD/1-/17	8	1.23
NE/2-/19	6	0.92
SD/2-PR/19	5	0.77
SD/3-PR/1-PD/1-/16	5	0.77
SD/3-PR/2-PD/1-/15	5	0.77
SD/6-PD/1-/14	5	0.77
NE/1-PD/1-/19	4	0.62
SD/1-NE/1-/19	4	0.62

One can also display the average duration of response for each response category either for the whole group (Figure 10.5) or by treatment groups, which will provide visual comparison of the treatment effects. As one can see from Figure 10.6, treatment A had longer duration for the better response categories and shorter duration for progression disease.

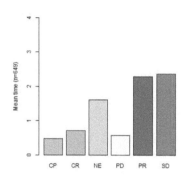 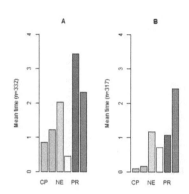

FIGURE 10.5
All Groups.

FIGURE 10.6
By Group.

10.3   Transition of Response States

An important information about response profiles is the transition rate between each pair of states (s_i, s_j), i.e., the probability to switch at a given time position from state s_i to state s_j.

Mathematically, let $n_t(s_i)$ be the number of sequences that do not end in time position t and with state s_i at time position t, and let $n_{t,t+1}(s_i, s_j)$ be the number of sequences with state s_i at time position t and state s_j at time position $t+1$. The "1-step" transition rate $p(s_j|s_i)$ between states s_i and s_j is defined as

$$p(s_j|s_i) = \frac{\sum_{t=1}^{L-1} n_{t,t+1}(s_i, s_j)}{\sum_{t=1}^{L-1} n_t(s_i)} \tag{10.1}$$

with L the maximal observed sequence length.

The rates are assumed time position independent and the outcome is a stochastic Markov process matrix where each row i gives a transition distribution from the originating state s_i in t to the states in $t+1$. Transition rates provide information about the state changes observed in the data together with, on the diagonal, an assessment of the stability of each state. Similar transition matrix can be constructed for multiple time steps.

The following are the transition matrices for Treatments A and B. One can clearly see the difference of treatment effect, which indicates the superiority of Treatment A to B when comparing the probabilities of each pair of cells at the same location of their respective matrix.

```
computing transition rates for states CP/CR/NE/PD/PR/SD ...
(Treatment A)

          [-> CP] [-> CR] [-> NE] [-> PD] [-> PR] [-> SD]
[CP ->]    0.82    0.04    0.10    0.03    0.00    0.00
[CR ->]    0.00    0.86    0.11    0.03    0.00    0.00
[NE ->]    0.05    0.07    0.52    0.03    0.19    0.13
[PD ->]    0.00    0.00    0.00    1.00    0.00    0.00
[PR ->]    0.02    0.02    0.11    0.06    0.78    0.00
[SD ->]    0.01    0.00    0.10    0.05    0.22    0.61


computing transition rates for states CP/CR/NE/PD/PR/SD ...
(Treatment B)

          [-> CP] [-> CR] [-> NE] [-> PD] [-> PR] [-> SD]
[CP ->]    0.76    0.07    0.10    0.07    0.00    0.00
[CR ->]    0.00    0.76    0.18    0.06    0.00    0.00
[NE ->]    0.01    0.02    0.47    0.08    0.16    0.25
[PD ->]    0.00    0.00    0.00    1.00    0.00    0.00
[PR ->]    0.00    0.01    0.16    0.12    0.71    0.00
[SD ->]    0.01    0.00    0.11    0.16    0.07    0.64
```

10.4 Classification of Response Profiles

It is always with special interest in medical research to segment the patient population based on their responses so that the relationship between the responses and background data can be further investigated for each subgroup. For example, the recursive partitioning methods discussed in the previous chapter were especially useful for this purpose. For profile analysis, we need to define the similarities of response profiles via some appropriate metrics so that further analysis can be conducted.

10.4.1 Dissimilarities Between Response Profiles

Similar to the calculation of distance between two data points, one can define the distance between two response profiles. Let $A(x, y)$ be a count of common states between sequences x and y. Define a dissimilarity measure through the following general formula

$$d(x, y) = A(x, x) + A(y, y) - 2A(x, y) \tag{10.2}$$

Cluster Dendrogram

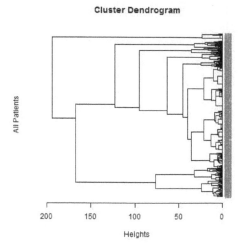

FIGURE 10.7
Clustering response profiles.

where $d(x, y)$ is the distance between sequences x and y. The dissimilarity is maximal when $A(x, y) = 0$. It is zero when the sequences are identical.

One commonly used method is, among others, the simple Hamming distance (Hamming [29]), which is the number of time positions at which two sequences of equal length differ. It can be defined as

$$HD(x, y) = l - A_H(x, y),$$

where $l = |x| = |y|$ is the common length of x and y, and $A_H(x, y)$ is the number of matching time positions. The Hamming distance with equation (10.2) by using $d(x, y)/2$ as proximity measure. With the defined distance between response profiles, cluster analysis was performed to aggregate the responses which have close proximity. The cluster dendrogram is graphically shown in Figure 10.7. Four clusters are formed by cutting the dendrogram with height of 100. One can further visualize the data distributions of these clusters with the multivariate methods such as multidimensional scaling.

10.4.2   Visualizing Clusters via Multidimensional Scaling

The determination of coordinates of multidimensional scaling can be briefly described as following. Let $\{x_1, x_2, \cdots, x_n\} \in R^d$ and the inter-point distance be $\delta_{rs} = ||x_r - x_s||$. Denote the dissimilarity matrix by $D = [(\delta_{rs})]$ and define matrix $A = [(a_{rs})]$ with $a_{rs} = -\delta_{rs}^2/2$. Define the matrix

$$B = [(b_rs)] = (I_n - 1_n 1_n'/n)A(I_n - 1_n 1_n'/n), \qquad (10.3)$$

where $b_{rs} = a_{rs} - \bar{a}_{r.} - \bar{a}_{.s} + \bar{a}_{..}$ One can extract the k largest positive eigenvalues $\gamma_1 > \gamma_2 > \cdots > \gamma_k$ of B with corresponding eigenvectors $Y_k = (y^{(1)}, y^{(2)}, \cdots, y^{(k)})$ that are normalized by $||y^{(j)}||^2 = \gamma_j$ for $j = 1, 2, \cdots, k$. The n rows of Y_k are the principal coordinates in k dimensions and can be plotted pairwise to visualize the data distributions.

The coordinates of the responses are plotted in Figures 10.8 to 10.11. Each cluster is highlighted in red color. One can easily see the different data distributions of these responses. Cluster 1 mostly aggregates at the upper-left corner of the plot and cluster-2 mostly aggregates around the center of the plot, while clusters 3 and 4 represent the responses not confirming with these two bigger groups. Since clusters 3 and 4 represent a small subgroup of patients in the analysis, we will focus our following discussions on clusters 1 and 2.

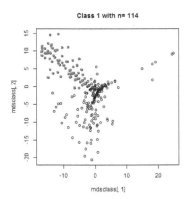

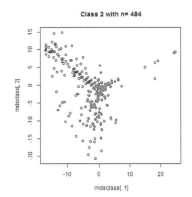

FIGURE 10.8
Cluster 1.

FIGURE 10.9
Cluster 2.

10.4.3 Response Profile Differences among Clusters

With further examination of the response profiles corresponding to the members of these clusters, interesting patterns of responses can be clearly observed, as shown in Figures 10.12 to 10.14. Cluster 1 consists of the patients who had disease status improved from Stable Disease to Partial Response or stayed at the same disease status. Cluster 2 consists of all the patients who had their disease status worsen from Stable Disease to Progressive Disease. It also includes the patients with disease status improved to Partial Response, nearly

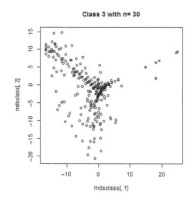

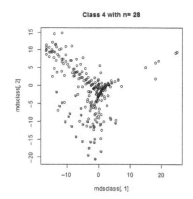

FIGURE 10.10
Cluster 3.

FIGURE 10.11
Cluster 4.

Complete Response, or Complete Response. Clusters 3 and 4 consist of patients who had improved disease status and none of the members in these clusters had any disease progression. Since any clustering method will be accompanied with different degrees of misclassification errors, it is likely some of the response profiles could have been better classified.

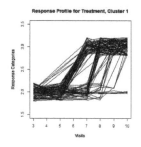

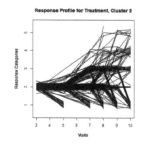

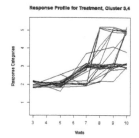

FIGURE 10.12
Cluster 1.

FIGURE 10.13
Cluster 2.

FIGURE 10.14
Clusters 3 and 4.

10.4.4 Significant Clinical Variables for Each Cluster

To further investigate the significant covariates which underlying the clusters, the patients' baseline covariates, the treatment received, and the laboratory data during the study are carefully examined. A regression analysis was conducted with the covariates, which was showing some degrees of significance after pre-screening. The results are shown below for clusters 1 and 2. Treatment **trtgrp** showed significance in cluster 2 as some patients in this cluster had substantial improvement into CR or PR, which is not the case for cluster 1. Treatment also showed a significant effect for clusters 3 and 4, however, since the clusters are small, the results were not showed here. Several laboratory tests and the abnormal serum protein levels show statistical significance in the model in addition to a few baseline covariates such as prior therapies received before the trial. Different sets of covariates appear in each cluster.

Analytical results for cluster 1:

```
Call: glm(formula = clustdata1[, 2] ~ n_prt + mmdur + strlab2 + AlbuminL +
     NeutrophilL + PlateletsH + RBCL + SerumIgAL + SerumIgAH +
     SerumIgGL + SerumIgGH + TotalProteinH, family = quasibinomial, data = clustdata1)

Coefficients:
             Estimate Std. Error t value Pr(>|t|)
(Intercept)   0.54234    0.35424   1.531 0.126264
n_prt         0.32146    0.09976   3.222 0.001335 **
mmdur         0.05649    0.03242   1.742 0.081933 .
strlab2      -0.90648    0.22449  -4.038 6.04e-05 ***
AlbuminL     -1.80328    0.54017  -3.338 0.000891 ***
NeutrophilL   2.00171    0.39783   5.032 6.32e-07 ***
PlateletsH    3.03917    1.31229   2.316 0.020876 *
RBCL         -0.95070    0.32570  -2.919 0.003635 **
SerumIgAL    -0.78516    0.28668  -2.739 0.006337 **
SerumIgAH    -0.88987    0.43682  -2.037 0.042045 *
SerumIgGL    -1.14937    0.34321  -3.349 0.000859 ***
SerumIgGH    -1.68069    0.54402  -3.089 0.002092 **
TotalProteinH -3.86644   1.24619  -3.103 0.002002 **
---
Signif. codes:  0 '***' 0.001 '**' 0.01 '*' 0.05 '.' 0.1 ' ' 1
(Dispersion parameter for quasibinomial family taken to be 0.7413571)
```

Analytical results for cluster 2:

```
Call: glm(formula = clustdata2[, 2] ~ trtgrp + prmel + AlbuminL + MonocytesH +
     RBCL + SerumIgAL + SerumIgAH + SerumIgGH, family = quasibinomial, data = clustdata2)

Coefficients:
            Estimate Std. Error t value Pr(>|t|)
(Intercept)  -1.5632     0.5387  -2.902 0.003837 **
trtgrp        0.7503     0.2818   2.662 0.007950 **
prmel        -0.8368     0.2506  -3.339 0.000888 ***
AlbuminL      1.5906     0.4943   3.218 0.001357 **
MonocytesH   -1.2791     0.6236  -2.051 0.040649 *
RBCL          1.0554     0.3597   2.934 0.003465 **
SerumIgAL     1.1015     0.3260   3.379 0.000772 ***
SerumIgAH     1.9697     0.4869   4.046 5.85e-05 ***
SerumIgGH     1.0588     0.3961   2.673 0.007707 **
---
Signif. codes:  0 '***' 0.001 '**' 0.01 '*' 0.05 '.' 0.1 ' ' 1
(Dispersion parameter for quasibinomial family taken to be 1.350058)
```

10.5   Modeling of Response Profiles via GEE

Generalized estimating equation (GEE) is a popular analytical method in the analysis of longitudinal data. Specifically, let $Y_{it} \in \{1, 2, \cdots, I > 2\}$ be the multinomial response for patient $i(i = 1, \cdots, N)$ at time $t(t = 1, \cdots, T_i)$ and assume missing data are missing completely at random (MCAR) as defined in Rubin (1976). Define $Y_{itj} = I(Y_{it} = j)$ for $j = 1, \cdots, I$, where $I(A)$ denotes the indicator function of the event A, and $Y_{it} = (Y_{it1}, ..., Y_{it(I-1)})'$ with the response category I omitted. Denote response vector by

$$Y_i = (Y'_{i1}, \cdots, Y'_{iT_i})'$$

and $(T_{i(I-1)} \times p)$ covariate matrix for patient i by $x_i = (x_{i1}, \cdots, x_{iT_i})'$. Let

$$\pi_{itj} = E(Y_{itj}|x_i) = Pr(Y_{itj} = 1|x_i), \quad \pi_{it} = (\pi_{it1}, \cdots, \pi_{it(I-1)})',$$

and

$$\pi_i = E(Y_i|x_i) = (\pi'_{i1}, \cdots, \pi'_{iT_i})'.$$

In addition, denote the link function by g and, for patient i at time t, is defined as

$$g[E(Y_{it}|x_i)] = g(\mu_{it}) = x_{it}\beta,$$

where β is the p-variate regression vector of interest.

10.5.1   Marginal Models

The choice of the link function, g, hence, the marginal model depends on the nature of the response scale. For ordinal multinomial responses, the family of cumulative link models

$$F^{-1}[P(Y_{it} \leq j|x_i)] = \beta_{0j} + \beta'_* x_{it}$$

or the adjacent categories logit model

$$\log(\pi_{itj}/\pi_{it(j+1)}) = \beta_{0j} + \beta'_* x_{it},$$

where F is the CDF of a continuous distribution and $\{\beta_{0j}, j = 1, 2, \cdots, J-1\}$ are the category-specific intercepts. For nominal multinomial responses, the baseline category logit model can be used

$$\log(\pi_{itj}/\pi_{itJ}) = \beta_{0j} + \beta'_j x_{it},$$

where β_j is the jth category-specific parameter vector. Note that the category-specific intercepts need to satisfy a monotonicity condition

$$\beta_{01} \leq \beta_{02}, \cdots, \leq \beta_{0(J-1)}$$

only when the family of cumulative link models is employed, and the regression parameter coefficients of the covariates x_{it} are category specific only in the marginal baseline category logit model.

10.5.2 Estimation of Marginal Regression Parameters

Let β be the p-variate parameter vector that includes all the regression parameters. Let the general estimating equations be

$$U(\beta, \hat{\alpha}) = (1/N) \sum_{i=1}^{N} D_i V_i^{-1}(Y_i - \pi_i) = 0, \qquad (10.4)$$

where $D_i = \partial \pi_i / \partial \beta$, $V_i(\beta, \hat{\alpha})$ is a $T(J-1) \times T(J-1)$ weight matrix, and α is the local odds ratio define next.

10.5.3 Local Odds Ratio

Consider the time-pairs $\{(1,2),(1,3),\cdots,(T-1,T)\}$. For each time-pair (t,t'), form an $J \times J$ contingency table such that the row totals correspond to the observed totals at time t and the column totals to the observed totals at time t'. Let $\theta_{tjt'j'}$ be the expected local odds ratio at cut point (j,j'), and $f_{tjt'j'}$ be the observed frequencies. Becker and Clogg (JASA,1989) showed that

$$\log(f_{tjt'j'}) = (\text{row, column, interaction effects}) \ + \ \phi^{(t,t')} \mu_j^{(t,t')} \mu_{j'}^{(t,t')},$$

and

$$\log(\theta_{tjt'j'}) = \phi^{(t,t')}(\mu_j^{(t,t')} - \mu_{j+1}^{(t,t')})(\mu_{j'}^{(t,t')} - \mu_{j'+1}^{(t,t')}).$$

Touloumis, Agresti, and Kateri (Biometrics, 2013) defined the parameter vector that contains the marginalized local odds ratios structure as

$$\alpha = (\theta_{1121}, \cdots, \theta_{1(J-1)2(J-1)}, \cdots, \theta_{(T-1)1T1}, \cdots, \theta_{(T-1)(J-1)T(J-1)})'$$

which can be estimated by MLE. For practical purpose, the score functions $\mu_j^{(t,t')}$ are simplified and let $\mu_j^{(t,t')} = j$, and the general association parameter $\phi^{(t,t')} = \phi$. Conditional on $\hat{\alpha}$ and the marginal model specification at times t and t' the probability

$$\{P(Y_{it} = j, Y_{it'} = j'|x_i) : t < t', j, j' = 1, \cdots, J-1\}$$

and V_i can be calculated, hence β can be estimated via the GEE (10.4).

10.5.4 Results of Modeling

Data from cluster 1 were analyzed using the method discussed above using programs in R and the results are shown in the following *estimates*. Similar analysis was also conducted for other clusters but are not shown here. The data beyond visit 10 were not included in the analysis. The beta level at baseline and the age both show significant effects, but the treatment only offers a moderate effect as patients from this cluster mostly only improved their disease status from SD to PR at the best. However, a similar analysis for cluster 2 shows significant treatment effect. Many patients in cluster 2 had improved their disease status from SD to PR, CP, and even CR.

Model and the parameter estimates:

```
GEE FOR ORDINAL MULTINOMIAL RESPONSES
Link : Cumulative logit

call: ordLORgee(formula = Resp ~ factor(vis_num) + factor(trtgrp) +
      age + beta + n_psct, data = xdat, id = pt, link = "logit",  LORstr = "category.exch")

Coefficients:
                    Estimate    san.se    san.z  Pr(>|san.z|)
beta01               1.35026   0.93392   1.4458      0.14824
beta02               3.77986   0.92989   4.0648        5e-05 ***
factor(vis_num)5    -0.18590   0.12986  -1.4315      0.15228
factor(vis_num)7    -1.67880   0.19489  -8.6143      < 2e-16 ***
factor(vis_num)8    -1.94960   0.30028  -6.4926      < 2e-16 ***
factor(vis_num)9    -2.65717   0.33162  -8.0128      < 2e-16 ***
factor(vis_num)10   -3.14579   0.36401  -8.6420      < 2e-16 ***
factor(trtgrp)2     -0.01723   0.24606  -0.0700      0.94418
age                 -0.03010   0.01399  -2.1513      0.03145 *
beta                -0.05015   0.05593  -0.8967      0.36986
n_psct              -0.25766   0.17849  -1.4435      0.14887
---
Signif. codes:  0 '***' 0.001 '**' 0.01 '*' 0.05 '.' 0.1 ' ' 1
```

The odds ratio of of the effect for each covariate:

```
                   Est. Coefficients    Odds Ratio of Better Responses
beta01                   1.35025563                         0.25917400
beta02                   3.77985846                         0.02282592
factor(vis_num)5        -0.18590278                         1.20430517
factor(vis_num)7        -1.67880421                         5.35914371
factor(vis_num)8        -1.94960456                         7.02590874
factor(vis_num)9        -2.65716906                        14.25587441
factor(vis_num)10       -3.14579299                        23.23809566
factor(trtgrp)2         -0.01722873                         1.01737800
age                     -0.03010472                         1.03056244
beta                    -0.05015157                         1.05143045
n_psct                  -0.25765765                         1.29389577
```

The correlation matrix of the parameters estimated in the final model:

```
                  beta01 beta02 (vis_5)(vis_7) (vis_8) (vis_9) (vis_10)(trtgrp)2 age    beta n_psct
beta01              0.87   0.85   0.00   0.00    0.00    0.01    -0.01   -0.03 -0.01     0  -0.09
beta02              0.85   0.86   0.00  -0.01   -0.04   -0.04    -0.06   -0.03 -0.01     0  -0.10
factor(vis_num)5    0.00   0.00   0.02   0.01    0.02    0.01     0.01    0.00  0.00     0   0.00
factor(vis_num)7    0.00  -0.01   0.01   0.04    0.03    0.03     0.03    0.00  0.00     0   0.00
factor(vis_num)8    0.00  -0.04   0.02   0.03    0.09    0.06     0.06    0.01  0.00     0   0.00
factor(vis_num)9    0.01  -0.04   0.01   0.03    0.06    0.11     0.07    0.00  0.00     0   0.01
factor(vis_num)10  -0.01  -0.06   0.01   0.03    0.06    0.07     0.13    0.01  0.00     0   0.02
factor(trtgrp)2    -0.03  -0.03   0.00   0.00    0.01    0.00     0.01    0.06  0.00     0   0.00
age                -0.01  -0.01   0.00   0.00    0.00    0.00     0.00    0.00  0.00     0   0.00
beta                0.00   0.00   0.00   0.00    0.00    0.00     0.00    0.00  0.00     0   0.00
n_psct             -0.09  -0.10   0.00   0.00    0.00    0.01     0.02    0.00  0.00     0   0.03
```

The estimated local odds ratios:

The local odds ratio matrix indicates the disease status improved from visit 3 to visit 5 is about 3.5 times, and it continued to improve for visit 7, but the effects started to decline after that.

```
       [,1]  [,2]  [,3]  [,4]  [,5]  [,6]  [,7]  [,8]  [,9] [,10] [,11] [,12]
 [1,] 0.000 0.000 3.525 3.525 4.534 4.534 1.437 1.437 1.250 1.250 0.887 0.887
 [2,] 0.000 0.000 3.525 3.525 4.534 4.534 1.437 1.437 1.250 1.250 0.887 0.887
 [3,] 3.525 3.525 0.000 0.000 2.512 2.512 3.422 3.422 1.982 1.982 1.424 1.424
 [4,] 3.525 3.525 0.000 0.000 2.512 2.512 3.422 3.422 1.982 1.982 1.424 1.424
 [5,] 4.534 4.534 2.512 2.512 0.000 0.000 1.383 1.383 1.347 1.347 1.062 1.062
 [6,] 4.534 4.534 2.512 2.512 0.000 0.000 1.383 1.383 1.347 1.347 1.062 1.062
 [7,] 1.437 1.437 3.422 3.422 1.383 1.383 0.000 0.000 1.591 1.591 1.375 1.375
 [8,] 1.437 1.437 3.422 3.422 1.383 1.383 0.000 0.000 1.591 1.591 1.375 1.375
 [9,] 1.250 1.250 1.982 1.982 1.347 1.347 1.591 1.591 0.000 0.000 1.604 1.604
[10,] 1.250 1.250 1.982 1.982 1.347 1.347 1.591 1.591 0.000 0.000 1.604 1.604
[11,] 0.887 0.887 1.424 1.424 1.062 1.062 1.375 1.375 1.604 1.604 0.000 0.000
[12,] 0.887 0.887 1.424 1.424 1.062 1.062 1.375 1.375 1.604 1.604 0.000 0.000
```

10.6   Summary

In this chapter, we conduct a case study of data from a good size clinical trial by studying the entire efficacy response profile of each patient and to explore the relationship between response profiles with covariates, which were collected either before or during the trial. Many graphical tools are used to visualize the data and make it easier for researchers to understand the data distribution and to compare the efficacy of treatments. Various analytical tools for multivariate data are also used to segment the patient population into subgroups so that the treatment effects can be further analyzed in each subgroup and possibly to discover better personalized treatments.

Medical research with dynamic treatment allocation designs is gaining popularity, especially for the early phase of drug discovery. Patients' responses are usually tracked at every cycle or visit in this type of study so that the proper treatment can be assigned for the next cycle of the trial to maximize the treatment benefit. Therefore, it is crucial to extend the conventional clinical data analysis, which mostly only emphasizes the end results to a broader scope covering more extended observations so that the analytical results can be more informative to both the clinical communities and the patients.

Bibliography

[1] Achim Zeileis, Torsten Hothorn, Kurt Hornik (2008) Model-Based Recursive Partitioning, *Journal of Computational and Graphical Statistics*, 17:2, 492-514.

[2] Agresti, A. (2002) Categorical Data Analysis. Wiley, 2nd Edition.

[3] Andrews, D.F., Gnanadesikan, R., Warner, J.L. (1973). Methods for assessing multivariate normality. In *Multivariate Analysis* III (P.R. Krishnaiah, ed.), Academic Press, New York, p.95-116.

[4] Archer, K.J. (2010) rpartOrdinal: An R Package for Deriving a Classification Tree for Predicting an Ordinal Response. *Journal of Statistical Software*, v.34(7), 1-17.

[5] Arpino, B., Cannas, M. (2016). Propensity score matching with clustered data. An application to the estimation of the impact of caesarean section on the Apgar score. *Statistics in Medicine*, v35, p2074-2091.

[6] Becker, R. A., Chambers, J. M. and Wilks, A. R. (1988) The New S Language. Wadsworth & Brooks/Cole (for S version).

[7] Becker, M. and Clogg, C. (1989). Analysis of sets of two-way contingency tables using association models. *Journal of the American Statistical Association*, 84, 142-151.

[8] Breiman, L. (2003). Manual—setting up, using and understanding random forests V4.0. Available at ftp://ftp.stat.berkeley.edu/pub/users/breiman/Using random forests v4.0.pdf.

[9] Breiman L., Friedman J. H., Olshen R. A., and Stone, C. J. (1984) Classification and Regression Trees. Wadsworth.

[10] Breiman, L. (2003). Manual—setting up, using and understanding random forests V4.0. Available at ftp://ftp.stat.berkeley.edu/pub/users/breiman/Using random forests v4.0.pdf.

[11] Buhlmann, P. and van de Geer, S. (2011) Statistics for High-Dimensional Data, Springer-Verlag, New York.

[12] Chambers, J. M., Cleveland,W. S., Kleiner, B., and Tukey, P. A. (1983) Graphical Methods for Data Analysis. Wadsworth & Brooks/Cole.

[13] Cohen A (1980). "On the Graphical Display of the Significant Components in a Two-Way Contingency Table." Communications in Statistics-Theory and Methods, A9, 1025-1041.

[14] Das, P., Roychudhury, S., Tripathy, S. (2018) sigFeature: Significant feature selection using SVM-RFE & t-statistic.

[15] Evers L., Messow C.M. (2008) Sparse kernel methods for high-dimensional survival data, *Bioinformatics*, 24 (14) p1632-1638.

[16] Emerson, J. D and Strenio, J. (1983). Boxplots and batch comparison. Chapter 3 of Understanding Robust and Exploratory Data Analysis, eds. D. C. Hoaglin, F. Mosteller and J. W. Tukey. Wiley

[17] Eric B. Holmgren (1995), The P-P Plot as a Method for Comparing Treatment Effects, *Journal of the American Statistical Association*, Vol. 90, p360-365.

[18] Fan, J. and Lv, J. (2008) Sure Independence Screening for Ultrahigh Dimensional Feature Space (with discussion). *Journal of Royal Statistical Society, B*, 70, 849-911.

[19] Fan,J. and Li,R. (2001) Variable selection via penalized likelihood. *J. Am. Stat. Assoc.*, v96, p1348-1360.

[20] Fang, Y. and Wang, J. (2012) Selection of the number of clusters via the bootstrap method. *Computational Statistics and Data Analysis*, 56, 468-477.

[21] Fisher, R.A. (1922). On the Mathematical Foundations of Theoretical Statistics, Philosophical Transactions of Royal Society of London, *Ser. A*, 222, 309-368.

[22] Florek, K., J. Lukaszewicz, J. Perkal, and S. Zubrzycki (1951). Sur la liaison et la division des points dun ensemble fini. *Colloquium Mathematicae* 2, 282-285.

[23] Fraley, C. and Raftery, A. E. (2002), Model-based clustering, discriminant analysis, and density estimation, *Journal of the American Statistical Association*, 97, 611-631.

[24] Freedman, D. and Diaconis, P. (1981) On the histogram as a density estimator: L_2 theory. *Zeitschrift für Wahrscheinlichkeitstheorie und verwandte Gebiete* 57, 453–476.

[25] Friendly, M. (1994). "Mosaic Displays for Multi-Way Contingency Tables." *Journal of the American Statistical Association* 89, 190-200.

[26] Gabadinho, A., Ritschard, G., Muller, N. and Studer, M. (2011). Analyzing and Visualizing State Sequences in R with TraMineR.

[27] Gnanadesikan, R. (1977). Methods for statistical data analysis of multivariate observations. John Wiley & Sons, New York.

[28] Gower, J. C. (1971). A general coefficient of similarity and some of its properties. *Biometrics*, 27, 857-874.

[29] Hamming RW (1950). Detecting and Error Correcting Codes, *Bell System Technical Journal*, 29, 147-160.

[30] Hansen, B. Full matching in an observational study of coaching for the SAT. *Journal of the American Statistical Association*. 2004; 99:609-619.

[31] Hartigan JA, Kleiner B (1984). "A Mosaic of Television Ratings." *The American Statistician*, 38, 32-35.

[32] Hastie, T., Tibshirani, R. and Friedman, J. (2011). The Elements of Statistical Learning, Data Mining, Inference, and Prediction.

[33] Hennig, C. (2007) Cluster-wise assessment of cluster stability. *Computational Statistics and Data Analysis*, 52, 258-271.

[34] Ho, D., Imai, K., King, G., Stuart, E. (2011) MatchIt: Nonparametric Preprocessing for Parametric Causal Inference. *Journal of Statistical Software*, June 2011, Volume 42, Issue 8.

[35] Hoerl, A. and Kennard, R. (1988) Ridge regression. *In Encyclopedia of Statistical Sciences*, vol. 8, pp. 129-136. New York: Wiley.

[36] Hothorn T, Hornik K, Zeileis A. Unbiased recursive partitioning: a conditional inference framework. *J Comput Graph Stat*. 2006;15(3):651-674.

[37] https://www.stat.berkeley.edu/ breiman/RandomForests/cc_home.htm

[38] Huber, P. (1997). Speculations on the path of Statistics. In: *The Practice of Data Analysis: Essays in Honor of John W. Tukey*. Princeton University Press, Princeton, NJ.

[39] Imbens, G. and Rubin, D. (2015) Causal inference, Cambridge University Press.

[40] Ishwaran, H., Kogalur, U., Blackstone, E. and Lauer, M. (2008) Random survival forests. *The Annals of Applied Statistics*, Vol. 2, No. 3, p841-860.

[41] Kalbfleisch, J., Prentice, R. (2002). The Statistical Analysis of Failure Time Data. Wiley series in probability and statistics.

[42] Kaufman L, Rousseeuw PJ (1987) Clustering by means of medoids. In: Dodge Y (ed) Statistical Data Analysis Based on the L_1 Norm and Related Methods, pp 405-416

[43] Kaufman L, Rousseeuw PJ (1990) Finding Groups in Data: An Introduction to Cluster Analysis. John Wiley & Sons.

[44] Khan F.M., Zubek V.B. (2008) Support vector regression for censored data (SVRc): a novel tool for survival analysis, in: F. Giannotti, D. Gunopulos, F. Turini, C. Zaniolo, N. Ramakrishnan, X. Wu (Eds.), Proceedings of the Eighth IEEE International Conference on Data Mining (ICDM), IEEE computer society, California.

[45] Landwehr, J.M., Pregibon, D., and Shoemaker, A.C. Graphical Methods for Assessing Logistic Regression Models. *Journal of the American Statistical Association* 1984; **79**: 61-71.

[46] Leisch, F. (1999) Bagged clustering. Working Paper 51, SFB "Adaptive Information Systems and Modeling in Economics and Management Science".

[47] Light, R.J. & Margolin, B.H. (1971). An analysis of variance for categorical data. *JASA*, v.66, p.534-544.

[48] Litvakov, B.M. (1966) On an iterative method in the problem of approximation of a function. Automat. Remote. Contr. (USSR) 27 (4)

[49] Liu, J., Ji, S., Ye, J. (2009). SLEP: Sparse Learning with Efficient Projections, Arizona State University. URL: http://www.public.asu.edu/jye02/Software/SLEP.

[50] Madan Gopal Kundu, and Jaroslaw Harezlak (2019) Regression trees for longitudinal data with baseline covariates, Biostatistics & Epidemiology, 3:1, 1-22,

[51] Mallows, C.L. (1998). The zeroth problem. *Am. Stat.* 52, 1-9.

[52] McGill, R., Tukey, J. W. and Larsen, W. A. (1978) Variations of box plots. *The American Statistician* 32, 12–16.

[53] Meier, L., van de Geer, S., Bühlmann, P. (2008). The group lasso for logistic regression. *J. R. Stat. Soc. Ser.* B 70, 53–71.

[54] Meyer, D., Zeileis, A., and Hornik, K. (2006), The strucplot framework: Visualizing multi-way contingency tables with vcd. *Journal of Statistical Softwa* 17(3), 1-48. URL http://www.jstatsoft.org/v17/i03/ and available as vignette("strucplot").

[55] Mo, Q., Wang, S., Seshan, V., Olshen, A., Nikolaus Schultz, N., Chris Sander, C., Powers, R., Marc Ladanyi. M., Shen, R. (2013). Pattern discovery and cancer gene identification in integrated cancer genomic data. *PNAS*, vol. 110, p4245-4250.

[56] Mo, Q., Shen, R., Guo, C., Vannucci, M., Chan, K., Hilsenbeck, S. (2018). A fully Bayesian latent variable model for integrative clustering analysis of multi-type omics data. *Biostatistics*, v19, p. 71-86

[57] Monti, S., Tamayo, P., Mesirov, J., Golub, T. (2003) Consensus Clustering: A Resampling-Based Method for Class Discovery and Visualization of Gene Expression Microarray Data. *Machine Learning*, 52, 91-118.

[58] Murdoch, D.J. and Chow, E.D. (1996). A graphical display of large correlation matrices. *The American Statistician* 50, 178-180.

[59] Nesterov, Y. (2004). Introductory lectures on convex optimization: A basic course, Operations Research.

[60] Nesterov, Y. (2007). Gradient methods for minimizing composite objective function, Technical report, Technical Report, Center for Operations Research and Econometrics (CORE),Catholic University of Louvain.

[61] Nevo, D. & Ya'acov Ritov (2017) Identifying a Minimal Class of Models for High-dimensional Data. *Journal of Machine Learning Research* 18, p1-29.

[62] Pearl, J. (2018). The seven tools of causal inference with reflections on machine learning. Association for Computing Machinery, v1. https://doi.org/10.1145/

[63] Pearl, J., Mackenzie, D. (2018). The Book of Why: The New Science of Cause and Effect. Basic Books, New York.

[64] Petersen, M., van der Laan, M. (2014). Causal Models and Learning from Data: Integrating Causal Modeling and Statistical Estimation, *Epidemiology*. v25, p418-426

[65] Piccarreta R (2001). A New Measure of Nominal-Ordinal Association. *Journal of Applied Statistics*, 28, 107-120.

[66] Piccarreta R (2008). Classification Trees for Ordinal Variables. *Computational Statistics*, 23, 407-427.

[67] R Development Core Team (2011). R: A Language and Environment for Statistical Computing. R Foundation for Statistical Computing, Vienna, Austria. ISBN 3-900051-07-0, URL http: //www.R-project.org/.

[68] Rosenbaum, P.R. and Rubin, D.B. The central role of the propensity score in observational studies for causal effects. *Biometrika* 1983; **79**: 516-524.

[69] Rosenbaum, P. R. (1989) Optimal matching for observational studies. *J. Am. Statist. Ass.* 84,1024-1032.

[70] Rosenbaum, P. (1991), A Characterization of Optimal Designs for Observational Studies, *Journal of the Royal Statistical Society* 53, 597-610.

[71] Rosenbaum, P.R. *Observational Studies.* New York: Springer-Verlag 1995.

[72] F. Rosenblatt (1962), *Principles of Neurodynamics: Perceptron and Theory of Brain Mechanisms*, Spartan Books, Washington, DC.

[73] Rubin, D.B. Assignment to a treatment group on the basis of a covariate. *Journal of Educational Statistics* 1977; **2**: 1-26.

[74] Segal, Mark Robert. 1992. Tree-Structured Methods for Longitudinal Data. *Journal of the American Statistical Association* 87 (418). Taylor & Francis Group: 407-18.

[75] Sekhon, J.S. (2006). Alternative Balance Metrics for Bias Reduction in Matching Methods for Causal Inference. *Working Paper.* *http://sekhon.berkeley.edu/papers/SekhonBalanceMetrics.pdf* 2006.

[76] Sheather, S. J. and Jones M. C. (1991) A reliable data-based bandwidth selection method for kernel density estimation. *J. Roy. Statist. Soc.* B, 683-690.

[77] Shivaswamy P.K., Chu W., Jansche M. (2007) A support vector approach to censored targets, in: Proceedings of the 2007 Seventh IEEE International Conference on Data Mining (ICDM), IEEE Computer Society, California, 2007, pp. 655-660.

[78] Silverman, B. W. (1986) Density Estimation. London: Chapman and Hall. *Journal of the American Statistical Association* 87 (418). Taylor & Francis Group: 407-18.

[79] Simon N, Friedman J, Hastie T, Tibshirani R. (2013). A sparse-group lasso. *J Comput Graph Stat.* 22(2):231-45.

[80] Scott, D. W. (1979) On optimal and data-based histograms. *Biometrika* 66, 605–610.

[81] Scott, D. W. (1992) Multivariate Density Estimation. Theory, Practice and Visualization. New York: Wiley.

[82] Steinhaus, H. (1957). Sur la division des corps matriels en parties. *Bulletin of the Polish Academy of Sciences* 4, 801-804.

[83] Stuart, E. and Green, K. (2008). Using Full Matching to Estimate Causal Effects in Nonexperimental Studies: Examining the Relationship Between Adolescent Marijuana Use and Adult Outcomes. *Dev Psychol.* v44, p395-406.

[84] Sturges, H. A. (1926) The choice of a class interval. *Journal of the American Statistical Association*, 21, 65–66.

[85] Therneau, T.M. and Atkinson, E.J (1997) Introduction to Recursive Partitioning using the RPART Routines Mayo Foundation.

[86] Tibshirani, R. (1996). Regression analysis and selection via the Lasso. *Journal of the Royal Statistical Society Series* B 58 267-288.

[87] Tipping, M. (2001) Sparse Bayesian learning and the relevance vector machine. *Journal of Machine Learning Research* 1, 211-244

[88] Touloumis, A., Agresti, A. and Kateri, M. (2013). GEE for multinomial responses using a local odds ratios parameterization. *Biometrics*, 69, 633-640.

[89] Tsai, K.T. Assessing Regression Modeling with Ordinal Responses. *Presentation at the Joint Statistical Meetings of the American Statistical Association* 2008.

[90] Tukey, J.W. (1962). The future of data analysis. *Ann. Math. Statist.*, 33, 1-63.

[91] Tukey, J. W. (1977) Exploratory Data Analysis. Section 2C.

[92] Van Belle, V., Pelckmans, K., Suykens, J., Huffel, S. (2009) Learning transformation models for ranking and survival analysis. Technical report, 09-135, ESAT-SISTA, KULeuven (Leuven, Belgium)

[93] Van Belle, V., Pelckmans, K., Huffel, S., Suykens, J. (2009) Support vector methods for survival analysis: a comparison between ranking and regression approaches. Technical report, 09-235, ESAT-SISTA, KULeuven (Leuven, Belgium)

[94] Van Belle, V., Pelckmans, K., Huffel, S., Suykens, J. (2011) Improved performance on high-dimensional survival data by application of Survival-SVM, *Bioinformatics*, v 27, p87-94

[95] Van Belle, V., Pelckmans, K., Huffel, S., Suykens, J. (2007) Support Vector Machines for Survival Analysis., in: E. Ifeachor, A. Anastasiou (Eds.), Proceedings of the Third International Conference on Computational Intelligence in Medicine and Healthcare (CIMED).

[96] Velleman, P. F. and Hoaglin, D. C. (1981) Applications, Basics and Computing of Exploratory Data Analysis. Duxbury Press.

[97] Venables, W. N. and Ripley, B. D. (2002) Modern Applied Statistics with S. New York: Springer.

[98] Ward, J. H. (1963). Hierarchical grouping to optimize an objective function. *Journal of the American Statistical Association* 58, 236-244.

[99] Wilkerson, M.D., Hayes, D.N. (2010). ConsensusClusterPlus: a class discovery tool with confidence assessments and item tracking. Bioinformatics, 2010 Jun 15;26(12):1572-3.

[100] Yang, Y., Zou, H. (2014). A fast unified algorithm for solving group-lasso penalize learning problems. Stat Comput.

[101] Yu, Y. and Lambert, D. 1999. Fitting Trees to Functional Data, with an Application to Time-of-Day Patterns. Journal of Computational and Graphical Statistics 8 (4). Taylor & Francis Group: 749-62.

[102] Yuan, M., Lin, Y. (2006). Model selection and estimation in regression with grouped variables. *J. R. Stat. Soc. Ser.* B 68, 49–67.

[103] Zhang, H., Ahn, J., Lin, X., Park, C. (2006) Gene selection using support vector machines with non-convex penalty. Bioinformatics, v22, p88-95.

[104] Zou H, Hastie T (2005). Regularization and Variable Selection via the Elastic Net. *Journal of the Royal Statistical Society* B, 67(2), 301-320.

Index

Milton Keynes UK
Ingram Content Group UK Ltd.
UKHW030900141024
449569UK00007B/483